JACK FRENCH

PRIVATE EYELASHES

radio's lady detectives

JACK FRENCH

PRIVATE EYELASHES

radio's lady detectives

BearManor Media
2004

Private Eyelashes: Radio's Lady Detectives

Published in the USA by

BearManor Media
P. O. Box 750
Boalsburg, PA 16827

bearmanormedia.com

Cover design by Matthew French
www.designsupportsystems.com

Typesetting and layout by John Teehan

ISBN—1-9714570-8-5

Dedication

This book is dedicated to my wife, Cathy

And to my fellow radio historians:
John Dunning, Jim Cox, Elizabeth McLeod,
Jay Hickerson, Martin Grams, Jr., Thomas DeLong,
Jim Harmon, and Stewart Wright

Table of Contents

Foreword

by Barbara J. Watkins
(Host/Producer of "Don't Touch That Dial";
SPERDVAC columnist)

From the beginning, detective and mystery shows have been my favorite genres of radio programs. In the late 1970s when I started to collect the old-time radio shows that I had grown up listening to, I made a list of all known series. My collecting goal was to have at least one show of every series and as many as I could find of my favorites.

Chances are if I had been asked then to name a few of the detective series, I'd have mentioned *Sam Spade, Philip Marlowe, Richard Diamond*, and *Johnny Dollar*. If they were really interested in my answer I'd have added some of my personal favorites: *Boston Blackie, Nick Carter, The Fat Man, Crime Files of Flamond, Candy Matson*. But that would probably be the only female detective that came to mind. So how can someone write an entire book about the subject?

Well, Jack French has. He has brought to the foreground many of the ladies who worked alongside their more famous male partners, and those who worked alone, in relative obscurity and were long forgotten. We learn about who these women were and their places in the world of detection. And in most cases, who the actresses were who played them and some background on their creators.

We are treated to a fresh look at the wives whom we perhaps took for granted: Pam North, Nora Charles, and the secretaries/girl Fridays: Patsy Bowen, Margot Lane, Della Street. These and many others are covered in these pages. In addition to discussing their radio roles, Jack also covers how their characters differed from one medium to another as they moved among novel and radio and TV and film. (Patsy Bowen's steady boyfriend was Scubby? I never knew that!)

i

More exciting for me is reading about those lady sleuths who have long been on my list of "wanted" shows, without success: *The Affairs of Ann Scotland, Helen Holden, Government Girl,* Anne Rogers of *Hot Copy, Susan Bright, Detective.*

It is also a treat to be introduced to some whom I had not known existed, such as Irene Delroy, Marie Revell, and Debby Spencer. A few of these episodes have surfaced to give us a glimpse of the world of lady detectives in the 1930s. We learn the history of these shows.

Then there is Candy Matson! I believe it was sometime in the 1980s when Jack and I first compared lists of this show. I suspect it was she who first gave him the idea for this book. To my ears Candy was unique among female characters in radio. After years of listening to soap operas and other shows aimed at the housewife when home sick from school, and actually all of the nighttime shows, too, I was used to the woman-in-trouble and the man-to-the-rescue scenario. What a dramatic change when Candy was in a tight spot and one expected Lieutenant Mallard to come to her rescue. NO! Candy got out of the jam herself! It wasn't until Natalie Masters explained that the series was originally written for a man, her husband Monty, to play the role that this role reversal made sense.

Readers beware, while reading this book you may have an irresistible urge to listen to these programs that Jack describes so vividly. Be prepared with a stack of tapes or CDs and join these lady crime fighters while you peruse the pages of this book. Let us hope that many more of these shows will eventually turn up for our enjoyment. In the meantime, we all owe a debt of gratitude to Jack for bringing them once again to life.

Preface

Feminine sleuths and crime-fighters were far more numerous in mystery novels, comic books, and juvenile literature than they ever were on network radio dramatizations.

Anne Katherine Green is frequently called America's "Mother of Detective Fiction" since she published her earliest mystery in 1878. Green, from her studio in Buffalo, NY, wrote dozens of popular stories and books for over 40 years, several of them featuring the exploits of Amelia Butterworth, an amateur sleuth. However, the first American woman to pen a detective novel was actually Metta Fuller Victor; her book, *The Dead Letter*, written under the nom-de-plume, Seeley Regester, was published in 1867.

Agatha Christie introduced her detecting heroine, Miss Jane Marple, in the mystery entitled, *Murder at the Vicarage* in 1930. That same year, on this side of the Atlantic, teenaged Nancy Drew made her initial appearance in *The Secret of the Old Clock*, written by Mildred Wirt Benson, under the pseudonym of Carolyn Keene, assigned to her by the Stratemeyer syndicate.

The comic books introduced lady crime-fighters in the early 1940s. Although *Wonder Woman* (the 1941 creation of psychologist W. M. Marston and artist Harry Peter) is usually credited as the first comic book heroine, she was preceded by *Miss Fury*, who debuted four months earlier. The latter was created and drawn by Tarpe (June) Mills. Several other crime-busting women appeared in comic books during World War II, most of them drawn by lady artists. These included *Blonde Bomber and the Girl Commandos* (originally by Barbara Hall, and later by Jill Elgin), *The Black Cat* by Al Gabriele, *Yankee Girl* by Ann Brewster, *Miss America* by Pauline Loth, and *Toni Gayle* by Janice Valleau.

In 1943 Blue Ribbon Books published *The Great Women Detectives and Criminals*, edited by "Ellery Queen." Among the 22 short stories by

American and British authors, 18 of them featured a fictional lady sleuth. Among those detectives represented were: Elinor Vance (1924), Sally Cardiff (1934), Sally "Sherlock" Holmes (1937), Hildegarde Withers (1941) and Desdemona "Squeakie" Meadow (1943).

One of the first feminine crime-solvers to appear in film was *Miss Madelyn Mack*, portrayed by Alice Joyce, in a 1914 silent motion picture released by the Kalem Moving Picture Company. Miss Mack, the creation of author, Hugh C. Weir, was a virtual mirror image of Sherlock Holmes: a scientific, eccentric, brilliant detective, who performed classical music, analyzed tobacco ashes, and occasionally used a cocaine derivative.

While *Nancy Drew, Miss Marple,* and *Torchy Blane* (the latter created by Frederick Nebel) had no trouble leaping from the printed page to the silver screen, these three popular amateur sleuths never got their own radio shows. Most of radio's feminine crime-solvers originated with their broadcast creators, and had no direct roots in detective fiction, comic strips, or the movies. However, many of the wives and companions of male detectives, did originate in the crime fiction pages.

Some of radio's lady sleuths were employed as private investigators, including Ann Scotland, Candy Matson, and Phyl Coe. One network heroine was a lawyer, Marty Ellis Bryant, in *Defense Attorney* while another leading character, Mary Vance in *Miss Pinkerton, Incorporated* was both a law school graduate and a private detective. The other crime-solvers in this book had many occupations; they were intrepid reporters, secretaries, clerks, and office assistants, whose stock in trade was uncovering mysteries and stopping crime.

All of these women were smart, self-reliant, fearless, and with a few exceptions, they were all single. More importantly, these ladies, appearing at least a generation before "women's lib" came to prominence, were thoroughly independent. Unlike their female listeners in the radio audience, these women sleuths did not take orders from, but gave them to, their male companions. These lady crime-solvers were "take charge" women, so their boy friends, and other male associates, had no choice but to follow their lead.

The final selection of the feminine radio characters for this book was a difficult one, or as my former professors at the University of Wisconsin would term it, "the continual sifting and winnowing" process. To begin, every known lady detective who worked solo will be found within these pages, even when available historical data on her show is scarce. Also, in all the programs involving a husband and wife team, the better half will be discussed.

When we consider the secretaries, girl friends, and Gal Fridays of male detectives, the delineation of who's in and who's out becomes more complicated. In general, the parameters for inclusion in this book required that the female counterpart had to render significant assistance in mystery solving or crime-fighting. So Suzy, assistant to Dan Holiday in *Box 13,* and Effie, the secretary in *The Adventures of Sam Spade,* are not discussed in this book, as they seldom left the office. Some women associates of male investigators had their roles, and their contributions, increased or decreased over the years, depending on the scriptwriters. In these cases, I have tried to lean in the direction of inclusion.

Nikki Porter, who became the leading lady in *The Adventures of Ellery Queen,* when that "celebrated detective" jumped from the printed page to network radio, was a cordial and pleasant companion. However, Nikki never gathered a clue, interrogated a suspect, nor contributed to the solution of any investigation. Francis "Mike" Nevins, the co-author with Martin Grams, Jr. of the recent book, *The Sound of Detection; Ellery Queen's Adventures in Radio,* postulates that the format of Queen's radio series did not permit Nikki to participate in the deductive process.

Only American radio heroines are included in this book, so *Modesty Blaise* and other BBC heroines are absent. Also, lady sleuths represented only in an audition recording, such as Irene Tedrow's *Trouble is my Business* or Gloria Blondell's *Holiday Wilde* are not discussed in these pages. However, *Policewoman-USA* is included, even though the only episode is the audition disk, as it is related to a prior show, *Policewoman.* Not covered are lady spies or espionage agents, i.e. the female stars in *Top Secret* and *Man From G-2.* Juvenile leads, i.e. *Little Orphan Annie,* and teenaged sidekicks, such as Patsy Donovan and Joyce Ryan in *Captain Midnight,* are not included in this book, even though one could argue their roles involved significant crime-fighting. None of the lady sleuths in this book were created to attract a juvenile audience, with two exceptions: *The Lady in Blue,* and Barbara Sutton, companion of *The Black Hood.* Both characters originated in the comic books and, when brought to radio, were aimed at young listeners.

While I have been researching radio's lady detectives for nearly 20 years, it is likely that I have not yet found every one who appeared on network broadcasting. Also, reasonable people may disagree as to the exclusion of one of their favorite feminine counterparts on a male detective series. As all of us in the Old Time Radio community realize, our fact seeking process will never encompass all historical data. And so it is with my project. This book is not the end of my research into radio's lady detectives; it is merely one plateau. Please contact me through my publisher with any additions or corrections.

Acknowledgements

The Old Time Radio community is filled with caring, generous folks who willingly share their knowledge and archival material with their fellow researchers. Many of them are close friends of mine, but just as many of my helpers I have never met, including some overseas. During the past twenty years that I have been actively researching radio's lady detectives, I have been the beneficiary of much kindness from strangers around the globe, some of whom have sent audio copies of a previously uncirculated program, a pertinent clipping from an old magazine, or a lead to follow up on yet another feminine sleuth. To all of these wonderful people goes my sincere gratitude, for without their assistance and encouragement, this book would never have been completed.

Here are the contributors to my research, who are listed approximately in chronological order: Norman Cox, for information on *Meet Miss Sherlock* and *Susan Bright, Detective;* Peter E. Blau, a fellow Sherlockian, for transcript of NPR special; Bill Nadel, for voice identification on *It's A Crime, Mr. Collins* plus data on *Meet Miss Sherlock* and Dashiell Hammett; Jerry Austin, for data on *Phyl Coe Mysteries;* Stewart Wright, for review of *Candy Matson* scripts; Bud Cary, for background on Sondra Gair, Monty Margetts (now deceased) for her memories of *Meet Miss Sherlock;* Julian "Jay" Rendon, former *Candy Matson* soundman, for background on that series; Karen Plunkett-Powell, author of *The Nancy Drew Scrapbook,* for encouraging me to write this book; Dan Haefele, for suggesting the title of this book; Matthew French, my son, for designing the front cover; Lois Culver, widow of Howard, for recollections of *Defense Attorney;* Dick Olday, for copies of *Phyl Coe Mysteries* and *Dan Dunn;* Jim Harmon, for copies of *The Black Hood* and *Policewoman;* Dave Amaral, NBC engineer, for extensive data on San Francisco radio personalities and programs; Allen J. Hubin, for copies of *Crime on the Waterfront* and *Time For Love;* Larry Groebe, for script of *Hot Copy;* Francis "Mike" Nevins and Martin Grams, Jr. for background on Nikki Porter; Ron Sayles, for biographical data on OTR stars and copies of *McGarry and His Mouse;* Steve Lewis, bookseller, for data on *Abbott Mysteries;* Ken Klug, for background on *Two on a Clue;* Lynchburg College Library for loan of *Improbable Fiction;* and Ron Ramirez, Philco collector, for data on *Phyl Coe Mysteries.*

Other generous contributors were: Jerry Haendiges, for copies of *Two on a Clue;* Crinkly Bottom Books, for edition of *The Yellow Violet;* Henry Leff, aka "Mallard", for his recollections of *Candy Matson;* Dr. Robert Kiss of England, for biographical data on Dudley Manlove; Anthony Tollin, for data on Margot Lane; Derek Tague, for research in *Variety* and *Billboard;* Stephen Jansen, for copies of *Manhunt* and other series; Barbara Watkins, for editorial assistance and copies of *Michael & Kitty* and *Sara's Private Caper;* Bobb Lynes, for voice identification of Michael Piper; Newport News Public Library for loan of *The Quality of Mercy;* John Ruklick, for copies of *Lady of the Press, Stand By For Crime,* and other series; Michael Hayde, for data on *Dragnet;* Elliot Vittes, for background on script writer, Louis Vittes; Douglas Dahlgren, for recollections of Natalie Masters; Christian Blees of Germany, for data on Marlene Dietrich; Jim Johnson, my high school debate partner, for research at Museum of Television & Radio; Michael Dewees, for genealogical data on Monty and Natalie Masters; Barry R. Weiss, for script reviews of *Candy Matson* and *Sandra Martin ;* Laura Wagner and The Stumpf-Ohmart Collection, for loan of photographs; Ben Ohmart, indexer of this book; Steve Kelez, for copies of *Abbott Mysteries* and *Adventures of Leonidas Witherall;* Great American Radio, for copies of *Danger, Mr. Danfield;* and Jeanette Berard of Thousand Oaks Library, for background on *Candy Matson, Lady of the Press,* and *Crime Files of Flamond.*

In the last month before my manuscript deadline, the following people responded to my urgent requests: Dan Riedstra, for research in Chicago periodicals; Jim Cox, for information on Florence Freeman; Karl Schadow, for *Quick As a Flash* log and background about Margot Lane; Ted Davenport, for copies of *Front Page Farrell* and *Wendy Warren & the News;* Dick Gayley, for sheet music of "Candy;" Jamie Kelly and Ian Grieve of Australia, for data on their country's version of *It's A Crime, Mr. Collins;* Lucille Bliss, for recollections of San Francisco radio programs, and Dr. Glenn Smith and Professor Anthony "Tony" Maiello of Music Department at George Mason University for identifying theme music of *Phyl Coe Mysteries.*

And my special thanks to Michael Henry and the staff of the Library of American Broadcasting at the University of Maryland for all their assistance in making available their extensive archives of scripts, periodicals, and reference books.

The First Ladies

Before Sam Spade, Nick Carter, or Dick Tracy took to the airwaves, a number of radio's lady detectives had been solving mysterious crimes and bringing the guilty to justice. These feminine sleuths ran their own private investigative offices, gave orders to male subordinates, including their boyfriends, and handled the toughest of cases. In this group of five "first ladies," only one was married, Kitty Keene. The other four were single, self-reliant, and with the exception of Carolyn Day, beholden to no man.

IRENE DELROY

A Broadway singer and dancer of the 1920s, Irene Delroy became one of radio's first lady detectives in 1932 after her movie career stalled. She was born in Bloomington, IN in 1898 and by the time she was twenty, she was a musical ingenue on the Broadway stage. Delroy was in several productions of the *Ziegfeld Follies*, plus other Manhattan musicals, some with a teenaged Eleanor Powell, and the *Greenwich Village Follies* with Martha Graham. Her success and beauty elevated her to the cover of *Dance* magazine.

Delroy went to Hollywood in 1929 and over the next two years, starred in Vitaphone musical films churned out by Warner Brothers. In 1930 she was featured in *Divorce Among Friends, The Life of the Party,* and *Oh! Sailor, Behave!,* the latter with Charles King. However, after *Men of the Sky* (1931), with Jack Whiting, she had trouble finding work. Musicals were being scaled back due to declining customer enthusiasm, so Delroy, at Warner Brothers, and many other singers and dancers had their contracts terminated.

In 1932, Delroy turned, probably reluctantly, to radio. One of the largest companies producing radio transcription disks for syndicated marketing to small and medium-sized stations was Radio Transcription Company of America, Ltd., usually called "Transco." It was located in Hollywood, and according to radio historian, Elizabeth McLeod, had its own recording studio on Cosmo Street, plus it rented other nearby studios when business was good. Lindsay McHarrie, formerly a local radio production manager at KHJ, supervised most of Transco's radio shows, using performers, including Elvia Allman, and production personnel from KHJ.

From 1930 to 1938, Transco turned out a steady stream of modestly-budgeted radio series, usually adventure, mystery, or comedy shows. A full series could encompass as few as 13 episodes, or as many as 52, each of them 15 minutes in length. The standard episode would consist of 12 minutes of drama or comedy, with a minute and a half of recorded music at both the beginning and the end. This enabled the local station, if they could find paying sponsors, to have one of their staff announcers do three minutes of voice-over commercials.

Transco was interested more in quantity than quality so most of their series were less than impressive, including *Police Headquarters, Comedy Capers,* and *Hollywood Spotlight.* How much Delroy knew about the company, or what amount of money she was paid for her services, may never be determined by radio historians. But it is logical to assume that one of the deciding factors was that she was allowed to use her own name in her two syndicated series. Both Transco and Delroy must have enjoyed this arrangement since she got the benefit of the radio publicity and Transco had bagged its first known star.

Although Transco went out of business in 1938, many of its recordings were still in syndication in the 1940s and '50s, because during World War II, Bruce Eells Productions bought the rights to the entire catalog of defunct Transco, reissued the disks, and kept the more profitable programs in circulation for another two decades. Many of these disks are now in the hands of collectors and commercial audio companies, some of whom have begun to re-release them, either in a full series, or as individual episodes.

We know of at least two Transco series that Delroy starred in: *The Transcontinental Murder Mystery* and *The Stratosphere Murder Mystery.* The actual occupation of Irene Delroy, a skilled sleuth, was never made clear in either series, but the assumption would be that she was an independent investigator. The status of her two assistants, who helped her on both national crimes and classified government cases, was equally unclear since they had limited, or no, enforcement authority.

Her boyfriend, Jimmy Gifford, was a reporter for the vaguely named "Press Association." A policeman, Sergeant Fitzgerald, merely identified himself to their clients, with a casual, "I'm from the Department." Jimmy and Fitz, as the officer was usually called, got on each other's nerves regularly and this provided some humorous moments since Delroy liked both of them and she had to smooth out their disputes.

Whatever her position, Irene Delroy had absolute cooperation of every agency or person in authority that she met. The captain of a passenger ocean liner, when he learned of her mission, declared, "Every officer and crew member are at your service." Despite her discretion and confidential investigations, her reputation and fame always preceded her. In one episode, when she came to Chicago on a classified assignment from "the War Department," a dozen reporters were in her hotel lobby, trying to get an interview with the lady detective.

The initial episode of *Transcontinental Murder Mystery* began on the deck of *The Scottish Queen,* an ocean liner bound for the U.S. from Great Britain. In London, Delroy has just captured Edmund Thurmond, a master thief who has stolen $ 500,000, and she is taking her prisoner back to the States for trial. While the ship's captain is congratulating her on her triumph, Delroy receives a cablegram from Scotland Yard which advises that the missing half-million dollar loot is hidden somewhere on the ship. Moreover, several unknown criminals, posing as passengers, will stop at nothing to find the stolen bounty. By the end of the episode, Gifford has recognized Blackie Peroni (a gangster using a phony passport), Fitz has almost recovered from sea sickness, and an unknown assailant has just murdered the ship's captain in his stateroom.

Delroy's second series, also recorded in 1932 and aired through syndication beginning in 1933, was *The Stratosphere Murder Mystery.* The first episode starts with Fitz at the Chicago airport, meeting Delroy on the plane from Washington so he could take her to her hotel. Unknown to them, Giffford has piloted his own plane from Washington and he joins them in their hotel room. Delroy then explains her secret mission. She is designated by the government to protect the life of Navy Lieutenant Commander Christopher Markham, who is perfecting his latest invention. This new weapon of war, involving cosmic rays and described as "the most important discovery since gun powder," is scheduled to be tested during a flight to the upper stratosphere. But sinister forces are determined to kill Markham and steal his invention.

According to Gifford, one of the chief suspects is a former associate of Markham's, Dr. Wallace, who works in a nearby laboratory. He and his strange, crippled assistant, Laufler, are protected by half a dozen Negro guards. When told this, Sergeant Fitz responds, "Oh, a minstrel show." At this point, Markham arrives and briefs the trio on how he needs protection as he fears for his life. Before he can explain further, the sound of an explosion outside interrupts him. All four race to the windows and see the blazing residue of Markham's car. "That bomb was meant for me!" he solemnly states as the episode ends.

MARIE REVELL

While both of these series starring Delroy were of average quality, the writing, acting, and production values made them entertaining and worthy of syndication. However, that was not true of a third Transco series from that same period, *Nemesis, Inc.,* in which a character named "Marie Revell" was featured. The opening lines by the announcer were somewhat pretentious:

> ANNOUNCER: On the top floor of one of the largest buildings in the loop of Chicago is the office of the world's most unusual corporation. The name on the door of this office is calculated to fill the observer with wonder, and a chill of unpleasant speculation creeps up the spine of all who read the gold lettering: Nemesis, Incorporated. Within the office, a young and attractive woman is seated at a large desk. She is Marie Revell, Chicago's greatest private detective and the junior partner in this amazing organization.

Although none of the cast have yet been identified, the lead character was far from appealing since the woman playing Marie Revell sounded snippy, argumentative, and irritable. She treated clients and subordinates with the same annoying, curt attitude. The plot veered from logic on almost every page. Revell, who was in charge of a large detective agency, had virtually no employees. She received guidance from a voice on a filter mike, who may have been her dead father, calling himself "Nemesis" but this was not clear either. She answered all incoming telephone calls herself and treated prospective clients with disdain.

In the first episode, Revell is on the telephone, dictating a "Help Wanted" advertisement to *The Tribune* when Pete Smith, posing as a window washer, swings on a rope, and crashes through her top floor window. After obtaining first aid, he applies for the vacancy and is hired on the spot. With the below dialogue, it's difficult to tell if the scriptwriter was trying to be funny or was simply inept:

MARIA: I keep getting phone calls from Peggy Donnelly, the disgustingly rich importer's daughter.

PETE: I know about Ol' Donnelly; he owns all the tea in China, or something.

As the first episode continued, Peggy tries to convince Maria on the phone that her father has been killed, but the lady sleuth thinks it is merely a kidnapping and she formulates a plan to put Pete in a tuxedo and check out the local nightclubs for the missing man. She changes her mind seconds later when Pete finds the corpse of Mr. Donnelly, braced against their office door bell.

It is hard to imagine that the typical radio listener, even in the Great Depression, would have been anxious to hear the next episode.

PHYL COE

Phyllis Coe, called "Phyl" by her friends, according to her radio announcer, was, of course, sponsored by Philco. In the 1930s, it was rather commonplace for the radio sponsor to dictate the names of the main performers on a program. Harvester Cigars sponsored the comedy series, *Harv and Esther,* and the Natural Bridge Shoe Company had a 15-minute variety show in which the leads were called Nat and Bridget. Palmolive, which sold soap and beauty products, sponsored a one hour musical anthology; on this series, the leading male and female vocalists had to use the names Paul Oliver and Olive Palmer.

Phyl Coe Radio Mysteries was more than just a crime drama featuring a "beautiful girl detective," it was a quiz contest for its entire radio audience. Philco's advertising agency, Geare-Marston of Philadelphia, handled all the contest details and prizes. The announcer for the show explained in detail the contest procedures in the first episode, and thereafter, at the beginning of each quarter-hour program, provided a summary of the below:

ANNOUNCER: Here's your chance to play detective! Phyl Coe
Mystery's on the air. The new, thrilling, chilling
radio contest that pays you huge cash prizes for
having fun. Every week at this same time over this
station, your Philco radio tube dealer brings you,
transcribed, the Phyl Coe Mystery contest. $ 50,000
in cash for the winners. $ 50,000 in cash for solving
simple mysteries. Weekly prizes! Big grand prizes! A
real contest of skill. Anybody can enter. Nothing to
buy. And you may win. All you need is the free
Philco Mystery Book your Philco tube dealer will
give you. Tells you all the amazing details. Shows
you how easily you can win. Lists all the prizes.
Contains free entry blanks. You cannot win without
this book. Get your free copy right away. All ready?
Got your thinking caps on? Listen carefully. Keep
your wits about you. Better have paper and pencil
handy. Jot things down in your Philco Mystery
Book, of course, if you already have it. And if you
haven't, be sure to get it as soon as possible. Phyllis
Coe, her friends call her Phyl Coe, the beautiful girl
detective, daughter of the late Philip Coe, world
famous criminologist, is going to solve a mystery
right before your ears, and then you are going to
show how she solved it. C'mon, let's play detective!

There is conflicting data on the specific dates this series actually aired.
Jerry Austin, an old-time radio collector in California, has seen the original transcription disks, and the twelve existing shows are dated one week
apart, from February 29, 1936 to May 16, 1936. However, the September 15, 1937 issue of *Broadcasting* magazine contained a brief press release
from "Philco Radio & Television Corporation" stating that the *Phyl Coe
Radio Mystery* had just begun its first week and it was being broadcast by
239 radio stations throughout the U.S. The presumption would be that
the shows were originally recorded in 1936, but for some reason, not
aired until the following year.

In the thirties, Philco was a major player in the radio industry. Its
origins go back to 1892 with the formation of the Spencer Company in

Philadelphia, which manufactured and sold carbon arc lamps. By 1906, the firm had switched its primary production to storage batteries, and thirteen years later, the business changed its name to Philadelphia Storage Battery Company. In the 1920s, most radios sold could be powered by a storage battery, which was essential in many regions, usually rural, where homes did not yet have electricity. Under an abbreviated name of Philco, the company gradually moved into the manufacturing of radios and radio tubes. Later, the widespread popularity of its Cathedral Model 20 (nicknamed the Double X) catapulted Philco to the forefront in radio sales.

Our factual data about *Phyl Coe Radio Mysteries* is sketchy at best. No one has yet identified the woman who performed the title role. Bud Collyer played the co-lead of Tom Taylor; he was the boyfriend of Phyl Coe, a mystery novelist, and a college graduate of "dear ol' Montague." Collyer's father, Clayton Johnson Heermance, had the poor judgment to transfer his full name to his son, tack on "Junior," and then give him the traditional nickname of "Bud." As soon as he could, the son dropped out of Fordham Law School, borrowed his mother's maiden name of Collyer and became Clayton "Bud" Collyer, radio performer. Most vintage radio buffs remember him fondly as the voice of both Clark Kent and Superman on Mutual.

In addition to Collyer, there were two other men in the supporting cast of *Phyl Coe Radio Mystery* who were destined for long-term success behind the microphone; House Jameson and Jay Jostyn played different, recurring roles on this lady detective series. Jameson went on to major roles on many network shows, although he was probably best known as the long-suffering father on *The Aldrich Family*. Jostyn had leading parts on dozens of soap operas but found his ideal role starring in *Mr. District Attorney*.

Concerning the rest of the cast and crew, including the director, scriptwriter, and sound effects personnel, nothing has been documented. Since all three of the known actors in the series were working in the New York City confines at the time, a rebuttable presumption exists that the production facilities were somewhere in the Manhattan area. While the total number of episodes is not certain, there are twelve consecutive shows in circulation today.

The theme music for this series was a curious choice. It was "Procession of the Sardar" from the symphonic suite, *Caucasian Sketches,* written in 1895 by Russian composer, Mikhail Ippolitov-Ivanov. He lived most of his life in Moscow, composing many operas, suites, and chamber music pieces, and eventually became the conductor of the Bolshoi Theater. He died at age 76 in 1935, one year before his composition was recorded as the theme music of *Phyl Coe Mysteries,* and it's fairly certain that his estate never received any royalties for this usage.

In her series, Phyl Coe was alternately identified as either a detective or an amateur criminologist, as though the two titles were identical. In her various adventures, she either happened upon a crime and solved it on the spot, or was requested to assist a victimized business in uncovering the perpetrator of a mysterious felony. Usually the cases involved a homicide, and, with the exception of her first case (where a man was killed at close range with a pistol), the causes of death were rather unusual. One victim was dispatched by poison, injected via a hypodermic needle, another was bashed in the skull and tossed overboard, and a third had his brain pierced by a hat pin, inserted through his ear. Occasionally, the script writer dabbled in science fiction. Coe and Taylor once visited an inventor in the mountains who had perfected a death ray tube, described as a glass cylinder small enough to be hidden in the pocket of a dressing gown.

Other than murders, the investigations that Phyl Coe dealt with were major thefts. She solved cases involving: the burglary of a fur store in which expensive sable coats were purloined, the theft of the Faith Diamond, and the loss of a painting by Dutch artist Jan Vermeer from a local museum by a thief who substituted a copy. Whatever the crime, or however her services were brought to bear upon a case, Phyl Coe was a "take charge" lady. In the "Case of the Dead Magician," Coe and Taylor were part of a large audience attending the magic show of Umberto the Great. In the middle of the performance, Umberto was shot by an unknown assailant and crumpled to the stage floor. Above the hubbub of the astonished crowd, Phyl Coe's voice rang out and she took control of the crime scene, by jumping onto the stage.

PHYL: Will everyone please remain seated.

1st WOMAN: Look! Umberto! He's dead!

PHYL: Quiet, please. Everyone, be quiet.

2nd WOMAN: The magician has been murdered!

PHYL: We don't know he's been murdered, but we know he's been shot. Now, can all of you ushers and stagehands hear me?

(*Mumbled assent from several*)

PHYL:	My name is Phyllis Coe. I'm a criminologist, an amateur, but still I've had experience with these cases. I want you ushers and stagehands to guard all the exits. Don't let anyone leave the theatre. One of you please go out and get the police.
1st MAN:	Okay, I will.
PHYL:	Now, is there a doctor in the audience?
2nd MAN:	Yes, Miss Coe, I'm a doctor.
PHYL:	Please come up and examine this man.
2nd MAN:	Of course, immediately.
TAYLOR:	Phyl, Phyl…
PHYL:	What, Tom?
TAYLOR:	Are you sure you want to get mixed up in this?
PHYL:	Of course, I do. It's my duty.

Every case was resolved successfully by Miss Coe, and at the end of each 15-minute show, she had confronted the guilty party, and whether man or woman, they confirmed Coe's solution by confessing their crime. This was the cue for the announcer to encourage the listening audience to write, in the Philco contest book, the clues which they thought had led Coe to identify the evildoer. Since these clues were never revealed in the broadcasts, even after the individual contests were over, it can be a little frustrating for today's fans to listen to the mysteries now, with no way to verify their conjectures.

Philco, as the sponsor, was not content with just the mention of "Phyl Coe" at least a dozen times in each of the 15-minute programs. Sometimes the announcer would caution listeners that they should replace their radio tubes with new tubes from Philco to ensure "reception will be good and clear and you won't miss a single important detail of any mystery program." During other shows, the commercial plug would shamelessly be put in the mouth of a character in the crime drama. A typical example of this was heard in the episode entitled "Last Will and Testament." In it, a middle-aged business man was taking his leave of his host for the night and he excused himself as follows:

MR. BOSWORTH: I noticed you have one of those new Double X Philco radio sets in your sitting room. I've heard this new Philco gives excellent on foreign stations.

Of course, the most blatant (or clever, depending on your point of view) commercialization by Philco involved the title of one of the episodes, "The Case of the Double X Mystery." As previously stated, Double X was the nickname of Philco's Cathedral radio, Model 20. The scriptwriter obviously had trouble coming up with a scenario to match this title, and he finally wrote a mystery which took place in a hospital. Four adjacent rooms in that medical establishment bore letters, not room numbers, which were X, Y, Z , and Double X, where the crime took place.

Years after this series ended, three of the known actors in its cast remained successful in broadcasting, both in radio and later, television. Jameson also had several movie roles. Collyer was probably the most successful of the three in his television endeavors; he was the star of *Beat the Clock* and *To Tell the Truth*.

After many periods of declining revenues, Philco filed for bankruptcy in 1962 and was eventually purchased by the Ford Motor Company. The auto-maker, in turn, sold Philco to GTE/Sylvania in 1974. At present, whatever may be left of Philco is owned by its parent company, Nordyne Corporation, while the brand name of Philco still occasionally appears on portable tape decks and clock radios sold in discount chain stores.

Collyer was only 61 years old when he died in September 1969. Two years later, Jameson followed him in death at the age of 68. Only Jostyn made it to his seventies; he was 72 when he passed away in July 1977.

Perhaps some vintage radio historian will eventually uncover the identify of the woman who was the voice of esteemed lady criminologist, Phyl Coe, and put an end to her final mystery.

CAROLYN DAY

To vintage radio historians, *Carolyn Day, Detective* is one of those tantalizingly elusive programs that is, thus far, unwilling to release its background secrets. Little data has been uncovered about the production company or the identity of cast and crew. It was a syndicated show, but its actual broadcast dates are inconclusive, although everything about its production values suggests the mid to late 1930s. A total of ten audio copies currently exist and most are five minutes in length. A radio serial with five-minute episodes was not unknown in the Thirties, but rather uncommon, except for comedy shows.

Carolyn Day, Detective was more of a soap opera than an adventure drama, and the lead character spent more time as a damsel in distress than a detective dedicated to deduction. Jim Cox, a radio historian, has identified Mary McConnell as the woman who played this feminine sleuth, but we know nothing more about McConnell. The young lady in the radio serial became a private investigator when she inherited the detective agency of her father, Randolph Day, upon his death.

Her boyfriend was Lt. Larry Lubeck of the local homicide squad; he has been previously listed as Larry Bixby, but recent examination of audio copies with better sound quality have disclosed his correct name. Lubeck viewed Carolyn only as a prospective spouse and tried unsuccessfully to convince her to give up the detective business and marry him. The actor who played Lubeck is yet unknown, but Howard McNear portrayed the main gangster, who had the unlikely name of Northrop C. Anderson. McNear was on hundreds of radio programs, but his most remembered role was Doc Adams on *Gunsmoke*. He would become more famous in the 1960s, playing Floyd Lawson, the barber, on television's *Andy Griffith Show*.

In the first episode, Carolyn takes Larry to see her father's office, which she has redecorated since inheriting it. The police officer is very critical of the additions and changes, including net lace curtains, chintz coverings on the chairs, and floral prints. He further comments that it couldn't be an authentic detective agency office since it has no spittoon, but she corrects him, pointing out that the spittoon is on the table, filled with fresh flowers. Discouraged, Larry then returns to his pursuit of her hand in marriage, in his typical brusque fashion:

LARRY: I want you to get out of this business and marry me, Angel Face. When you get tired of playing cops and robbers, and want to play house, you know where to find me.

Since this was primarily a soap opera, the plot was a little silly, and since the program had no announcer, Carolyn Day served as both the lead and the narrator. Each episode began with her reciting the events which had recently transpired, to brief the radio audience on what they might have missed, a custom on most soap operas. At the beginning of the fourth episode, Carolyn told her listeners the following:

CAROLYN: Bugsy and Luke had orders to kidnap me, but when Bugsy found out I was Randy Day's daughter, he wanted to let me go because he liked my father, even though he was a detective. But Luke was determined to go through with the kidnapping. Bugsy jumped him and there was a fight, and Bugsy and I ran out of the office, followed by Luke, who had gotten Bugsy's gun. He caught us and took us to see his boss.

That summary would probably not have encouraged many listeners to stay tuned, and a few may have even rolled their eyes in rejection. By the end of the sixth episode, the pattern of the series had been developed. After Carolyn's update on story line, the program began where it left off on the previous show. In nearly every episode Carolyn was confronted by an armed thug, to which she responded with surprise, fright, and charm, usually in that order. Unexplained sounds, i.e. doors opening and closing, or a scuffle in another part of the building, by unseen and unknown persons, were a regular occurrence, much like a 1930s stage melodrama

In radio's tradition of both the soap operas and the juvenile adventure programs, each episode of *Carolyn Day, Detective* ended with what the scriptwriter hoped was a startling development which would pull the audience back for its resolution. But since the majority of episodes had stilted dialogue, unconvincing sound effects, and pervasive mediocrity in the plots, this could lead to nothing more than the failure of this syndicated series.

KITTY KEENE

Kitty Keene, Incorporated was the first, non-syndicated, lady detective show on network radio, debuting on CBS on December 13, 1937. It was a soap opera, with Tchaikovsky's romantic "None But the Lonely Heart," played on a violin, as its theme song, and the announcer extolling the benefits of using Dreft washday detergent. Since it was a soap opera, it took this feminine sleuth about eight to ten weeks to solve a crime or mystery. But the well-known Perry Mason, whose featured program was also a soap opera, took the same amount of time, and nobody questioned his competence.

The voice of Kitty Keene was supplied by three different women over the nearly four years it was on the air: Fran Carlon, Gail Henshaw, and Beverly Younger. All of them were very versatile and successful radio performers. Carlon was born in 1913 in Indiana, but raised in Scranton, Pennsylvania.. After a brief career on stage and in movies, she came to network radio and quickly became one of the busiest actresses behind the microphone. She played Lorelei on *Big Town* and had recurring roles on *Attorney at Law, David Harum, Young Widder Brown, Blackstone the Magician, Ma Perkins,* and many more. Carlon was a founding member of the radio performer's union, American Federation of Radio Artists (AFRA), and proudly held AFRA card # 98. With the demise of network radio in the 1960s, she transitioned easily into television and continued to work in the entertainment industry until her death at the age of 80.

The other two women may not have been on as many different shows as Carlon, but both were skilled in their craft. Gail Henshaw did a lot of other soap operas, including playing a friend of Mary Noble on *Backstage Wife* and recurring roles on *Guiding Light,* among other programs. Beverly Younger played on a variety of Chicago shows, from soap operas (*Ma Perkins*) to juvenile adventure programs (*The Silver Eagle.*) All three of these leads portrayed Kitty Keene as courageous, dedicated, and resourceful.

Kitty Keene, Incorporated started on CBS, later went to NBC in 1938, and finished its run with two years on Mutual. It had its time slot changed frequently, and depending on the year, or the network, was heard at 5:30 PM, 10:45 AM, 11:30 AM, and as early as 8:30 AM in 1940. Proctor and Gamble, through the familiar advertising firm of Blackett-Sample-Hummert, was the sponsor for the entire four years, primarily advertising Dreft detergent for both the first, and last, two minutes of every episode.

One can only speculate on the origin of this lady detective's name, but it is probably not just a coincidence that the creator of the series was Day Keene. Taking his surname and matching it with an alliterative first name seems very likely. Alliteration abounded in the names of radio shows of that era, especially the soaps: *Lora Lawton, Houseboat Hannah, Joyce Jordan, Manhattan Mother, Helping Hand, Hilltop House, Central City,* etc. Although Kitty remained a detective, more or less, for the full four years, the "Incorporated" was dropped from the title midway through the run.

Kitty was an unlikely person to run her own detective agency. She was a mature woman, with a husband, Charles, and a grown daughter, Jill, when the series started, and by the time it left the airwaves, she was a grandmother. This made her the only radio lady detective to reach this status. Her husband, a newspaper reporter, wanted to be the sole support of his family, and occasionally Kitty would close her office and try to be the simple housewife he preferred. But when crime raised its ugly head in her vicinity, Kitty resumed her sleuthing.

The scripts make interesting reading, and are an insight into the minds of Frank and Anne Hummert, whose soap opera assembly factory of unsung writers produced so many women's daily dramas; at the high water mark, 36 different programs of theirs were airing concurrently. To cue the actors to the precise emotion or attitude in nearly every line uttered, the writers inserted terse instructions in capital letters. Thus, in episode # 527, (from July 1938) where Kitty has closed her office at the request of her husband, we find a typical passage in the script to illustrate these actor instructions.

KITTY: It'll make Charles happier to see me as a domesticated, placid little housefrau. (SINCERE) I'll start crocheting a luncheon cloth tomorrow.

CLARA: (AMUSED LAUGH)

KITTY: (PUZZLED) What's funny about that?

CLARA: (LITTLE LAUGH) You don't crochet luncheon cloths, Kitty. I think you better stick to detecting.

KITTY: (WOULD) I'd like to. You know what my work has meant to me.

CLARA: (DOES) Yes, I do. And I don't think Charles is being exactly fair in asking you to give it up.

KITTY: (FLYING TO HIS DEFENSE) Why shouldn't he, if he wants me to. He's my husband, isn't he?

CLARA: Well, don't snap my head off. That man can beat you for all of me. (AMUSED) Not that there's much danger of that.

KITTY: It seems life is just one problem after another. (THOUGHTFUL) We solve one, and—

CLARA: (INTERESTED) What was this story that Charles fell down on, Kitty?

And it was not just the overseers at the Hummert factory that paid very special attention to every word on every page. The sponsor, Proctor and Gamble, probably thinking that its two commercials were the most important part of each script, assigned a reviewer to go over the two or three pages at the beginning and end of every episode that contained the praises of Dreft. In 1938, this reviewer, "Mr. Bannvart," made minor changes to virtually every commercial, which were conveyed to the cast and crew via a special Hummert form entitled "Alteration."

One might think that with all the energy expended in the creation and oversight of these commercials, some deathless prose might have resulted. But, alas, those Dreft commercials were fairly routine stuff, sprinkled with occasional surprises. For example, in a program airing on the Fourth of July, housewives were advised that using Dreft would be a patriotic (and historic) gesture:

ANNOUNCER: Today is Independence Day! Celebrate it by washing the new, modern way. When you wash your fine fabrics, stockings and dainty underthings, switch to Dreft, the new miracle suds discovery that brings you all the advantages of the finest soaps—but none of soap's disadvantages! Dreft is the most remarkable washing suds ever made! Nothing like it has ever been offered to woman since the beginning of time!

If announcer Jack Brinkley could keep a straight face emoting the above into the microphone to his thousands of radio listeners, he probably had no trouble either describing a promotional premium that *Kitty Keene, Inc.* offered its fans in mid-1938. Since this was a detective series, it would be logical to offer a replica badge, or perhaps a pocket-sized magnifying glass. However, since the vast majority of its radio listeners were women, the premium offered was a four-section tube of makeup by the makers of Tattoo Lipstick. It had been woven into the story line in previous episodes, before Brinkley announced it was available to any Dreft user who had a dime in her pocket. Here's a summary of the "grand news."

ANNOUNCER: Say, we've got grand news today! You remember Kitty and Jill were so delighted with the new Purse Make-up Kit that Jill discovered? The makers of the famous Tattoo lipstick...designed a beautiful black and white case that looks like a big lipstick. In fact, that's what Jill thought it was! But lo and behold, instead of coming apart in two, this case unscrewed into four sections! Each section held a generous supply of a different Tattoo Cosmetic: lip rouge…face powder…all purpose cream…and Lip Youth…You know what it will mean to you to turn yourself out fresh and blooming at a moment's notice! When Kitty and Jill were so enthusiastic about their new kits, Procter and Gamble decided to make it available to all their friends!…Here's all you have to do to get it! Just send one Dreft box-top, and ten cents in coin, together with your name and address, to Dreft…D–R–E–F–T, Cincinnati, Ohio…Your Tattoo Kit will be forwarded postpaid in about two weeks!

Racketeers of every stripe seemed to occupy most of the crimes that Kitty was called upon to put right. Sometimes these criminal activities affected those closest to her, including her husband, her daughter, Jill, or her son-in-law, Bob Jones, a police lieutenant. These family members were occasionally victimized by evildoers, or in some instances, framed by them. Only four audio episodes of *Kitty Keene* have survived and all are from the year 1939. In one of them, dated April 8th, a grand jury is about to indict Jill for helping Alibi Ramoni's gang in a scheme involving a mortgage company where Jill was the accountant. Kitty dramatically rebukes the grand jury and insists that she be allowed to defend her daughter. The speech that follows is part lecture and part autobiography.

KITTY :   You all know I don't make statements I can't support. I've been fighting this case, trying to smash Ramoni's gang, while you gentlemen on the grand jury, by your own admission, were totally unaware of what was going on in your own city. Now, I don't very often talk about what's happened to me during any case, but this is an exception. Those few of you who don't know, may be interested to learn that I'm the same Kitty Keene who exposed and smashed the marriage racket, and who, in some measure, helped States Attorney

Thatcher break up the Juan Powers and Toohey gang. And…going back still further than that, I'm an ex-Follies girl. Oh, I've done things which I deeply regret. But I've never done anything, which when looked at from all sides, I need be ashamed.

If we can believe her statements (and she assures us that we can) then this lady was no ordinary private investigator; she was a crime solver of epic accomplishments. And although she had to utilize three different networks during her reign on the air, she still has the distinction of having one of the longest series for a woman detective in the Golden Age of Radio.

In the Driver's Seat

While most of the woman crime solvers on network radio were the wife, secretary or Gal-Friday of a male shamus, several ladies ran their own detective business, law office, or police division. While most of them had a boyfriend, he stayed in her shadow and never challenged her leadership. Mary Sullivan of *Policewoman* was based upon a widow's true story of her career in the New York Police Department and was on her own. *The Lady in Blue* had a leading lady crime-fighter whose only assistant was another woman. In addition to the five female private investigators in chapter one, there were seven other ladies who were in total charge of their careers.

Ann Scotland

Since neither audio copies nor original scripts have yet been discovered, *The Affairs of Ann Scotland* is another one of those network detective series that radio historians know a great deal more about the star than the character she portrayed. Arlene Francis played the title lead in this ABC show, which was broadcast from Los Angeles, CA. The half-hour show was sponsored by the Richard Hudnut Company and aired weekly on Wednesday evenings from October 1946 to October 1947.

Arlene Francis Kazanjian, who later created her professional name by dropping her surname, was born in Boston, MA on October 20, 1907. When she became famous, she started adjusting her date of birth and therefore many reference books list it as 1908, 1910 or other years. Arlene grew up in the Bronx section of New York and acquired her love of acting while still in grade school. Her bearded grandfather was a retired actor who had minor roles in touring Shakespearean productions, and his favorite pastime was to listen to Arlene recite poetry while he sipped his favorite booze.

Her father, a portrait photographer, was Greek Orthodox but he entered her in Mount St. Vincent high school, where he assumed the Catholic nuns would discourage her acting aspirations. Later, he was mortified to discover that, not only did the school have a drama society, but that his daughter flourished on their stage.

After graduation from high school, she got braces to correct a gap in her front teeth and was able to convince her mother to take her to Hollywood to check out her acting prospects. A California friend of her mother's was able to get Arlene an interview with David O. Selznick. Although he did not like the braces, nor the shape of her nose, he was sufficiently impressed with her potential. Shortly afterwards, she was cast in a walk-on role in *Murders in the Rue Morgue*, securing her first motion picture credit:

A Prostitute: Arlene Kazanjian

While she only had a bit part in that motion picture, she was selected for its publicity posters, which featured a risqué shot of her, clutching a nightgown, torn from her shoulder. Back in New York City, her father, passing Loews' State Theater, saw the movie poster, declared her "almost nude," and fired off a telegram ordering his wife to return immediately to New York with their fallen angel. Upon return, he placed her in Finch's Finishing School and prayed that she'd give up acting and stop embarrassing the family.

When Arlene completed finishing school, her father's next step was to attempt to make her into an entrepreneur by setting her up in a gift shop in uptown Manhattan. Her little business establishment occupied the first floor, while her father organized a photography studio above. Unknown to Papa Kazanjian, most of her "customers" were out-of-work actors, who congregated in her shop to escape the New York City weather elements.

One of these actors, impressed with Arlene's gift of mimicry, sent her to a radio audition for a job that required unusual voices parts, including animal sounds. She was offered the job and her father allowed he to accept it, since in his mind, radio wasn't really acting, it was just advertising in the air. He was also favorably influenced by the fact that her radio job paid nearly $100 a week, which was more than she earned in her retail establishment. As Arlene's radio drama career flourished in the next few years, both parents were proud of the fact that she was not really acting, since her work did not involve the movies or the Broadway stage, both of which were considered by them as vulgar.

Her mainstay in the 1930s were soap operas and she worked regularly on *Betty and Bob, Central City, Portia Faces Life, Second Husband* and was briefly on *The Rise of the Goldbergs*. Sometimes she was on a half dozen different soap operas the same day. But her intelligence, quick wit, and vivacious personality took her out of the soap opera regimen and brought her to the microphones of unscripted, audience participation programs and quiz shows. Arlene was the co-emcee with Budd Horlick on *What's My Line,* became the host of *Blind Date,* as well as *The Hour of Charm,* which featured Phil Spitalny's "All-Girl Orchestra."

Despite her substantial success in network broadcasting, Arlene still longed for the stage, and with or without her parents' consent, started getting small roles on Broadway. Her first leading role on Broadway was in *All That Glitters,* produced by George Abbott, in which she played a prostitute with a Spanish accent, who charmed her way into New York high society. Her next major role was in Maxwell Anderson's drama, *Journey to Jerusalem,* and doubtless her parents were ecstatic that she played the Virgin Mary and not a hooker again. Although she continued to make more money in radio, and later television, the live stage remained her favorite.

In 1942, the strain of overwork, and the pain of a crumbling marriage, caused a nervous breakdown, and after many weeks in a Manhattan area sanitarium. Arlene regained her mental health. She separated from her husband, Neil Agnew, moved to her own apartment, and continued her acting career on her own. Three years later, she got a divorce in Mexico from her husband, freeing her to marry fellow actor-turned-producer, Martin Gabel, with whom she had been carrying on a long term, not-so-secret romance.

After her second marriage in 1946, the couple relocated to Los Angeles, California, because most of Gabel's production obligations were on the West Coast. For a while, she was very uncomfortable, living in a region where star status was based upon one's affiliation to the motion picture business, and no one seemed to care one iota about her previous stage and radio success, or her cameo role in the film, *Stage Door Canteen.* Of course, soon she would go on to become one of television's luminaries, which meant a lot more to the Hollywood business associates of her husband.

Arlene was noticeably pregnant with their child when her agent called to say she was wanted for the title lead in a new West Coast radio series, playing a character she had never done before, a private detective. The series was called *The Affairs of Ann Scotland*, but by the time Arlene got around to writing her autobiography in 1978, she apparently had forgotten the exact title. She referred to it, not once but twice, in her book, as *The Adventures of Ann Scotland*, and her co-writer of that autobiography, Florence Rome, didn't bother to verify it either. Recalling her reaction about getting the radio series, Arlene wrote:

> "So I was happily back at the microphone. My burgeoning belly didn't matter at all, since I had to convey everything with my voice. So what if I had to waddle into and out of the studio? What if I had to stand sideways at the microphone? I was back in action and delighted about it.
>
> "I could have been called Scotland Yard, because that's how wide I was. I could hardly get close enough to the microphone to speak into it, but my voice still dripped with the breathy nuance of 'C'm up and see me sometime' variety. At one point, my friend Claire Trevor, listening to the show in New York, sent me a wire saying: 'You don't sound the least bit pregnant, but you sound as though you might be any moment!'"

Her series was a rarity for a network detective show, perhaps even a first, since three women were its driving force. In addition to Arlene in the lead, ex-movie actress Helen Mack directed the show and Barbara Owens wrote the scripts. Mack had moved from behind the radio microphone to the director's booth in the early 1940s. As a performer in soap operas, Mack was talented enough to take over the co-lead in *Myrt and Marge* when Donna Damerel suddenly died in February 1941. Later Mack segued into directing duties and was usually in charge of crime dramas (*The Saint*) or variety-comedies (*Alan Young Show* and *Meet Corliss Archer.*) Ken Niles was her first choice as announcer for *The Affairs of Ann Scotland* for he was the announcer on three other series she directed: *The Beulah Show, A Date With Judy,* and *The Amazing Mrs. Danberry.*

Obviously, the name of Ann Scotland was created to suggest crime-solving through investigative brilliance, as exemplified in the tradition of Scotland Yard. The popular name of Scotland Yard is that of the Metropolitan Police Force of London, England. The term originated from its headquarters' location during the period 1829 to 1890, which was then in an area named for a palace in which Scottish kings resided during their

London visits centuries prior. In 1890, the headquarters was moved to new buildings near the Thames, which, in turn, were extended to other nearby buildings, now called New Scotland Yard.

In another portion of her 1978 autobiography, Arlene spoke of her Ann Scotland character as "…a sexy girl detective…whose gimmick was that she trapped her quarry with guile and feminine wiles."

This series must have been a superb one, and for the missing audio copies, more's the pity. Backing up the talent at the top was a supporting cast of some of the best radio actors in Hollywood, a regular who's who of leading characters on other network shows: Howard Duff, Howard McNear, Cathy Lewis, and David Ellis. Musical themes and bridges were supplied by a topnotch organist, Del Castillo.

Interestingly, Francis had two appearances as a "guest detective" on *Quick As a Flash,* the popular game show. She appeared as herself on the December 10, 1944 show, but in the guise of Ann Scotland on their October 22, 1949 program.

After the end of this series in late 1947, Arlene continued her broadcasting success, becoming one of the busiest TV personalities in the 1950s. She was so popular that *Newsweek* had her on their July 19, 1954 cover. During one period, she had featured roles on three different television networks; she hosted NBC's *Home* five days a week, was emcee weekly on ABC's *Talent Patrol,* and was a panelist on *What's My Line* on CBS. The latter program lasted for 25 years, but long runs for her were not uncommon. Arlene's daily interview show for WOR radio began in 1963 and ran for 23 years; at its end, she was nearly 80. Her second husband, Martin Gabel, died in 1986. She lived to be 93 and had Alzheimer's disease when she passed away on May 31, 2001 in San Francisco, California.

DIANE LaVOLTA

Portraying a thinly-veiled composite of her silver screen persona, Marlene Dietrich launched the character of Diane (pronounced "Dee-on," not "Dy-an") LaVolta in January 1953 with her CBS radio series, *Time For Love.* This program was a reincarnation of her prior network series, for ABC, entitled *Cafe Istanbul,* which had gone off the air in December 1952. In this earlier series, Dietrich played Mlle. Madou, a singer and hostess in a Turkish cabaret. The name "Madou" was borrowed from the novel, *Arc de Triomphe,* written by Erich-Maria Remarque, a former lover of Dietrich's. *Cafe Istanbul* derived its theme and flavor from two films, Dietrich's *The Blue Angel* and Humphrey Bogart's *Casablanca*, so it mixed romance, music and foreign intrigue. Dietrich sang at least one song in each 30-minute episode, leaving most of the action to the local prefect of Turkish police.

The film *Casablanca* was based upon an unproduced play, *Everybody Comes to Rick's,* which was written by Murray Burnett and Jone Allison. The rights to this drama were purchased by Warner Brothers in 1941 for $ 20,000 and the resultant film went on to become a classic. In the 1950s, when Dietrich asked her friend, Burnett, to create a radio series for her, he drew heavily upon *Everybody Comes to Rick's* for the background of *Cafe Istanbul.* Incidentally, Burnett's original play was finally put on stage in 1991 in London, but the reviews were tepid ("...it's not *Casablanca....*") and it closed within a month.

When she switched networks in January 1953, Dietrich did more than just change the names of the show and its main characters. *Time for Love* cut down on the singing, even though Diane LaVolta used the cover of singer to disguise her secret identity as some sort of government agent. Just what her official position was, and with which country (U.S. or Great Britain?) remained deliberately vague in the scripts. When her American boyfriend, Michael Victor, complained he had lost her in England for two weeks, she coyly explained: "I had a little job to do for the Navy." Her never-defined authority to investigate and solve crimes may have been akin to that of Mrs. Emma Peel, the associate of John Steed, in television's superb sixties series, *The Avengers.* However, some authorities disagree on LaVolta's status; Christian Blees, a radio researcher in Germany, is convinced that this female crime solver merely found crimes by accident or was called upon for assistance by friends. Blees believes that LaVolta was neither a government agent nor an actual detective.

Another significant change in the second series was the abolition of a permanent location for the leading lady. No longer anchored to an Istanbul nightclub, her character meandered around the exotic locales of four continents. As a result, the voice of Bob Readick (playing Michael Victor) stood out in most episodes since he was the only character without a foreign accent. Readick missed several shows because of illness, and another actor, tentatively identified as Bert Cullen, took over for him. Despite the series title, LaVolta was so busy fighting crime and combating spies, she had little "time for love," if the complaints of Michael Victor can be believed.

Radio work certainly represented the tiniest portion of Dietrich's show business career, since the bulk of her earnings and fame derived from her appearances on the stage, in motion pictures, and in cabarets. Most of her biographies do not even mention her two network radio series. But in the early 1950s, she needed the money she got from radio. Her last film, *Rancho Notorious,* had tanked at the box office and movie offers had dried up. She refused to appear on television, insisting that people must pay in order to see her. So at this point in her career, network radio would have to do.

Dietrich and her press agents did all they could to obscure the details of her birth and early career abroad. She usually claimed to have been born in 1904 in Weimar, Germany, although in 1964 her actual birth certificate was located, establishing that Maria Magdalene Dietrich was born December 27, 1901 in Berlin.

She was the daughter of a Prussian army officer, who had resigned to become a police official. As a child, she was trained as a violinist, and eventually played in theaters for silent movies. Later, she gave up the violin, apparently because of a wrist injury, and she became a Berlin chorus girl around 1920, taking the first name of "Marlene." Dietrich first appeared in film in 1922 and then alternated between film and the stage in Berlin for the next four years. She had a lovely singing voice which she utilized well when the "talkies" began in Germany around 1929. It was about this time she also became proficient in playing the musical saw, which she used in her cabaret acts. Many years later, in World War II, she entertained American GI's not only with her singing, but also playing the musical saw.

Dietrich was a successful leading lady in European cinema by 1930 when she was brought to Hollywood by film director, Joseph von Sternberg. She rapidly rose as a box office star in the U.S. and became the legend she would embrace until her death, over fifty years later. Although urged by Nazi power brokers to return to Germany in 1937, she refused so Hitler banned all her movies in German territory. She not only became a U.S. naturalized citizen in 1939, she also went on tours to sell war bonds, entertained our troops abroad with the USO, and broadcasted anti-Nazi propaganda over the radio in her native tongue. As one of her biographers, Alexander Walker, wrote in 1984: " No female star of today would dare to personify the romantic side of the warrior myth as Dietrich (did) in image, words, and person." When she died in 1992, she had been retired for many years, but her name and image were still imprinted in the minds of many of her fellow citizens in her adopted country.

Dietrich's partner at the radio microphone in *Time for Love*, playing her lover, Michael, was Bob Readick, a fellow who grew up in radio. He had begun as a child actor on the popular children's show, *Let's Pretend*. When his voice changed, he moved on to adult roles in a variety of network programs. He was in the cast of the mystery series, *The Adventures of Father Brown*, and a Navy drama, *Now   Hear This*. Readick had leading roles on many radio soap operas, including *Young Dr. Malone, Rosemary*, and *The Second Mrs. Burton*. However, he is best known to American radio fans as one of the last, and maybe the best, of the actors to play *Johnny Dollar*, a role he took over in 1960.

In the surviving copies of *Time for Love's* episodes, Dietrich's German accent and Readick's standard American dialect were surrounded by a Tower of Babel supporting cast who supplied a myriad of foreign-flavored English. Among the regulars, who could produce any accent the script called for, were Guy Repp, Joe DeSantis, and Luis Van Rooten. Of the total 58 episodes of *Time For Love* that aired, only four audio copies are circulating in the U.S. However, 45 shows on original disks are archived in the Marlene Dietrich collection in Berlin, Germany. According to Christian Blees, some of these have been reissued recently by an European audio book publisher, with some additional narration in German.

LaVolta's missions, official and otherwise, took her to solve foreign mysteries and confront international criminals throughout the various regions of four different continents: Europe, Asia, South America, and Africa. In one adventure, she escaped a bomb planted at an airport in Algiers and then proceeded to break up a local counterfeiting ring "near the Casbah." Another case resulted in her being dispatched from London to a jungle in Tanganyika where she uncovered a gun smuggling operation. Her other cases involved the kidnapping of a bullfighter in Spain, a husband trying to murder his wife in Brazil, illegal drug dealing in Egypt, a homicide in Holland, and diamond smuggling in Greece.

But despite the constantly changing locales and crimes, several scripts sound as though they were cut from the same pattern. Many of her adventures begin with a chance meeting of LaVolta and her American lover in some exotic city. After a furtive attempt at romance, they separate and travel to their respective assignments (he was an author of crime novels) and soon she is in the midst of some sinister criminal plot. All of the characters, whether evil or well-intentioned, regularly furnished their obligatory observations on LaVolta's beauty, her revealing attire, and her sensual personality. LaVolta protected her vocal instrument by smoking her own special brand of cigarettes. "You know, my voice," she chastised any admirer of hers who offered her harsh, ordinary cigarettes. In addition to the theme song, *Time For Love,* Dietrich also sang another song in about a dozen of the extant episodes, usually one of her standards ("Lili Marlene," "Falling in Love Again," etc.) either in German or English.

When each story began to bristle with danger, LaVolta's courage and unabashed derring-do was demonstrated. In one scene, she went, unarmed, to look for evidence in the unlighted basement of a murder suspect. Another episode, which took place "in darkest Africa," an errant shot from her safari companion merely wounded a lion, so LaVolta grabbed a rifle and started into the tall grass to dispatch the enraged beast.

One would assume that this fearless and confident lady investigator would require no last minute assistance from any man. However, the ending for several episodes had a distinct resemblance to the action movie serials of the 1930s and '40s, where the hero arrived in a nick of time to save his beautiful feminine counterpart. No matter what final calamity endangered her, Michael Victor could reappear and save LaVolta from almost certain death. That left time for one last embrace and then a fade to commercial.

In all her cases, LaVolta mixed a liberal amount of charm and guile into her detective work. As a confirmed do-gooder, she was as interested in convincing the bad guy to repent as she was in gathering the evidence and slapping the cuffs on his wrists.

Time for Love aired weekly from January 1953 to May 1954, except for the traditional summer break from June to September. The series began as a sustaining (not sponsored) program, but Jergens took over the sponsorship in September 1953 and touted the praises of its hand cream until the series went off the air ten months later. The original radio contract was for 78 episodes, but Dietrich completed only 58, and the series ended by mutual agreement of the star and the sponsor. The cancellation of the show was because her agent had booked her into The Sahara in Las Vegas, Nevada, beginning December, 1953, for which she would earn an enormous fee. To prepare for this nightclub appearance, she needed time, so she began recording three radio shows in one evening. The sponsor argued about the scheduling and the scripts, and finally terminated her contract, which may have been what Dietrich wanted in the first place.

The next ten years were fairly successful for Dietrich. She took her cabaret act to Europe, South America, and the Middle East and was acclaimed for her performances. Her strong film roles in *Witness for the Prosecution* (1957) and *Judgment at Nuremberg* (1961) merited global praise. But the 1970s were cruel for her as age and injuries took their toll; she suffered several broken bones, falling on unfamiliar cabaret stages as her sense of balance deteriorated. American singers and comedians began to caricature her and her songs so she retreated into relative isolation in Europe. She made a brief appearance in a 1978 West German film *Just A Gigolo,* but remained aloof from most of her friends and fans. Dietrich was 90 years old and alone when she died in May 1992 in a small, untidy Paris apartment where she had spent the last years of her life as a friendless recluse.

MARTHA ELLIS BRYANT

A lady lawyer who spent virtually no time in any courtroom, preferring to stop crime and achieve justice through her own investigations, Martha ("Marty") Ellis Bryant was ABC's answer to Perry Mason. Screen star Mercedes McCambridge portrayed this mystery solving attorney who unmasked the guilty in the series, *Defense Attorney.* The program began on NBC, under a slightly different title, *The Defense Rests,* and while the audition show of April 17, 1951 is in circulation, it's unclear how many times, if any, that NBC aired their version of it.

The show on NBC, produced and directed by Warren Lewis with scripts by Cameron Blake, created Marty Bryant as an attorney in private practice who solved crimes, whether she had a client or not. Her very helpful boyfriend, Judson "Jud" Cramer, worked for a large daily newspaper. The opening, which Mercedes shared with announcer Don Stanley, was as follows:

BRYANT: I petition, I attest, I insist, that my client was wrongly and unjustly convicted of murder.

               No matter how long it takes, I'm going to prove it. It will be then, and only then, that the defense rests!

SOUND: Musical sting on organ

Mercedes McCambridge, *Defense Attorney* (*The Stumpf-Ohmart Collection*)

ANNOUNCER: The National Broadcasting Company is proud
to present Miss Mercedes McCambridge in *The
Defense Rests,* a new and exciting series of cases
from the files of an outstanding woman attorney,
Martha Ellis Bryant!

Mercedes McCambridge (*The Stumpf-Ohmart Collection*)

SOUND: Musical sting on organ

ANNOUNCER: When young, attractive Martha Ellis Bryant
                        chose law as a career, she was expected to take
                        advantage of her prominent family background

and attend to the minor legal needs of the rich. Instead she accepted the challenge of defending the defenseless.

While both the NBC and ABC versions were adventure drama, the writers frequently injected humor into most scripts. In an NBC episode airing April 10, 1952, Jud parodied the ABC opening by telling Bryant:

JUD: When Jud Barnes chose a lawyer as a girlfriend, he accepted the challenge of defending the defenseless girlfriend.

The ABC network version began August 31, 1951, under the direction of Dwight Hauser, a man with long and varied experience in radio. He was an actor in the supporting cast of *Dark Venture* and he handled sound effects for *The Hermit's Cave,* when it was resurrected on the West Coast. Later, he concentrated on directing; he was at the helm of *The Adventures of Bill Lance, The Man from Homicide,* and *I Fly Anything,* which starred Dick Haymes in a non-singing role. The name of Bryant's constant companion was changed from Jud Cramer to Jud Barnes, a reporter for *The Dispatch,* and Hauser cast his good friend and fishing buddy, Howard Culver, in that role. Two months prior, Culver had lost his job as the voice of *Straight Arrow* when that show was canceled. Joel Murcott, another pal of Hauser's, and Bill Johnson were brought on the team to write the scripts.

The series was produced in a Hollywood studio that ABC rented, on Vine Street, just below Hollywood Boulevard. With McCambridge (friends called her "Mercy") and Culver in the leads, Hauser relied on the following actors as his supporting cast: Tony Barrett, Harry Bartell, Dallas McKennon, Irene Tedrow, and Parley Baer. The show's music was composed and played by Rex Koury, who was then only a year away from creating the magnificent theme for radio's *Gunsmoke.*

Lois Culver, the widow of Howard, recently recalled the congeniality of the cast and crew. She would accompany Howard to Hollywood on the days he worked on *Defense Attorney* and she would run errands while he was at the microphone. During rehearsal breaks, Lois joined the cast who had "headed for the nearest ice cream parlor...and we sat around and talked...It really was an ice cream parlor. It was interesting listening to Mercy, who was quite a character. During that time she was involved in the political campaign and often held forth on that subject."

Of course, when ABC changed the title of the series from *The Defense Rests* to *Defense Attorney*, the opening had to change as well. Also, ABC had signed up two advertisers, both of whom felt that their products should be included in the program's introduction. As we all know, the sponsor is always right. so the new opening was:

ANNOUNCER: The makers of Kix, tasty, crispy, corn puffs, food for action, and the makers of Clorets, the new chlorophyll chewing gum that makes your breath "kissing sweet" present: Defense Attorney!

BRYANT: Ladies and Gentlemen: To depend upon your judgment, and to fulfill mine own obligation, I submit the facts, fully aware of my responsibility to my client, and to you, as defense attorney.

SOUND: Military music theme

Both network versions did everything they could to favorably promote their leading lady, Mercedes McCambridge, who at the time, worked for several movie studios. The NBC debut program of April 1951 included a message from the announcer, publicizing her three current films: *Inside Straight* (MGM), *Lightning Strikes Twice* (Warner Brothers) and *The Scarf* (United Artists). In the closing minutes of ABC's program of April 10, 1952, the network turned the microphone over to Betty Mills, an editor with *Radio Mirror Magazine,* who personally presented McCambridge with their annual award as radio's favorite dramatic actress.

Carlotta Agnes Mercedes McCambridge was born on March 17, St. Patrick's Day, 1918, which must have delighted her Catholic parents. She attended Catholic grade school, was then briefly in a convent, and later attended Mundelein College in Chicago. While a sophomore there, she auditioned for NBC on a lark and got several parts on soap operas almost immediately. Within five years, she was one of the busiest radio performers in the Midwest. Orson Welles, who later costarred with her on *Ford Theater,* magnanimously called her "the world's greatest living radio actress." Obviously, Welles was exaggerating for he certainly knew that Agnes Moorehead and Florence Williams, among others, were just as skilled as McCambridge, but Welles never let the facts get in the way of a good sound bite.

McCambridge's broadcasting success led to Broadway roles in the 1940s. An open call for actresses resulted in her being cast in her first film, 1949's *All The King's Men,* for which she would win an Academy Award for Best Supporting Actress. But her life was not all artistic triumph and professional success, as she related, with blistering candor, in her 1981 autobiography, *The Quality of Mercy.*

The book is not an easy read, since it is a stream of consciousness, with accounts of undated events, flashbacks, and haphazard chronology. Starting most chapters is akin to boarding a carousel while it is in motion. But Mercedes McCambridge was diligently honest and she discussed all the failures and contradictions of her maverick career and lifestyle. Raised a Catholic, she bragged of "lying to the Pope." In her autobiography she also confessed to her two failed marriages , her attempted suicide and her bouts with alcoholism, which for periods, derailed her acting career.

As one would expect, most of her memories of her entertainment career dealt with the stage and the silver screen. But, while she made no mention of *Defense Attorney* or Howard Culver in her book, she did devote several pages to her experience in radio. Her Chicago soap operas (*Guiding Light, Midstream,* etc.) were discussed, as were her West Coast programs (*I Love A Mystery, Abie's Irish Rose,* and *The Bob Hope Show*). She told an amusing tale of being cast in *Red Ryder.* The rehearsals were uneventful but just as the "On The Air" sign lit up, the boy who was Li'l Beaver ran to the bathroom. McCambridge played both her assigned role and the part of the Navajo lad on the show until the little actor returned.

In a tribute to the sounds effects personnel in the Thirties, she writes of a *Lights Out* episode in which she costarred with Boris Karloff. He played a vampire who killed her by biting through the skeletal frame of her spinal cord. NBC soundman, Tommy Horan, worked all week, trying unsuccessfully to create the sound of a spinal column being crushed. Finally inspiration struck; he put some hard mints in his mouth, placed his lips near the microphone, and crunched the mints with his teeth. The result was realistically horrific.

McCambridge discovered late in her career that her fans had more distinct memories of her at the radio microphone than they did of her movie, television or stage appearances. She wrote:

"People all over the country have retained for so long and so vividly the visions they themselves created merely from sounds," she later wrote.

"They don't remember what they saw on TV night before last, but they remember forever what they heard on radio. Radio was the best."

There are a total of seven extant episodes, one of the NBC version and six of ABC's. In most of them, Bryant reassured herself and her clients, "Conducting an investigation is decidedly out of my line; I'm an attorney, not a private detective." She then rushed to a crime scene and began her investigation. With the exception of the standard opening (which reminded some listeners of *Mr. District Attorney*) she seldom saw the inside of a courtroom, but spent most of her time grilling suspects in their lair, examining evidence for clues, and ultimately solving the crime. Even Jud reminded her occasionally, "Your business is defending, not apprehending." However, neither her resolutions, nor Jud's warnings, could restrict her to the duties of a practicing lawyer.

Most of the cases Bryant handled were homicides, and she was not reluctant to tackle old crimes still unsolved by the cold case squad. In one episode, she proved that a petty racketeer, who was serving a life sentence for a murder conviction 12 years prior, was innocent of the killing and she obtained his release. Another adventure concerned her efforts on behalf of a shopkeeper, on death row after his murder conviction months before, and Bryant not only got the principal prosecution witness to recant, she also identified the real killer.

But many of her crime sleuthing concentrated on recent homicides. When a teenaged friend of Jud's was accused in the hit-and-run death of a longshoreman, Bryant's speedy investigation cleared the boy and directly led the police to arrest the victim's supervisor, who had actually driven the car. In another case, a family dispute over a will resulted in the murder of both a son and his father, the latter death had been disguised as a suicide. Our lady sleuth, with a little help from Jud, not only proved the suicide was first-degree murder, but also her adroit questioning in the police station of the second son forced his confession to both killings.

Jud Barnes occupied more time handling assignments from his girlfriend than he did from his supervisor at the newspaper office, but if the administrators at *The Dispatch* noticed his priorities, they never complained. Barnes was dispatched on fact gathering missions routinely by Bryant, day or night, and he only rarely complained..."Whatever happened to the evenings we spent necking?" He was very useful to this female Perry Mason since his newspaper credentials apparently gave him access to the records at the local police, Department of Motor Vehicles, and other agencies that usually restricted any sharing of their official files.

During the time that this radio series aired, 1951-1952, McCambridge had branched out briefly into music and had recorded a few songs. The writers of *Defense Attorney*, certainly with her permission, worked one of these songs into the episode that aired August 31, 1951. It was relatively uncomplicated to introduce the recording into the show for Marty and Jud frequently took their breaks from crime-solving by eating burgers at "The Steak House." where a jukebox was usually playing. In this episode, when Jud heard a female voice coming from the jukebox, he broke off from their discussion of the murder mystery and the following dialogue ensued:

JUD:	Hey, listen to that tune on the jukebox. Do you know who's doing the vocal?
BRYANT:	Uh-uh.
JUD:	Mercedes McCambridge.
BRYANT:	You were expecting maybe Mario Lanza?
JUD:	I never knew she was a singer.
BRYANT:	If she is, she needs more than that record to prove it to me.

After the series ended in December 1952, Howard Culver continued working in radio (including roles on *Ft. Laramie*) and television, where he had parts in *The Virginian*, *Dragnet*, and *The Untouchables*. He was seen in dozens of *Gunsmoke* episodes, usually as the hotel clerk, and he also appeared in several now-forgotten movies, including *The Swarm*, *Desk Sergeant*, and *Code Red*. In the summer of 1984, he became critically ill following an ocean cruise to the Orient with his wife, Lois. Culver did not recover and he died on August 14, 1984; he was only 66 years old.

Mercedes McCambridge continued to get fewer, but probably better, roles in movies. She was in the casts of *Suddenly Last Summer* (1959), *A Touch of Evil* (which reunited her with Orson Welles in 1958), *A Farewell To Arms* (1957)and 1956's *Giant*. In the latter, she portrayed the feisty sister of Rock Hudson, for which she received an Oscar nomination for Best Supporting Actress. On the theatrical stage, she played Anne Sullivan in *The Miracle Worker*, Martha in *Who's Afraid of Virginia Woolf?* and she garnered a Tony nomination for her performance in *Lone Suicide at Schofield Barracks*. As of this writing, she is in her mid-80s and retired.

THE LADY IN BLUE

In the spring of 1951, a lady crime-fighter was transported from the comic book pages to the radio airwaves, taking only enough time enroute to change the color of her costume. While it has not been proven conclusively that NBC's *The Lady in Blue* was a direct steal from the comic book heroine, *Lady Luck,* the circumstantial evidence is very compelling.

Will Eisner, a native New Yorker, gave up his original goal of being a stage designer, and went on to great success as a writer and illustrator in the comic book business. In 1940, he created an action hero named "The Spirit" whose exploits were promulgated in a 16-page comic section, syndicated in the Sunday supplements of several newspapers. "The Spirit," who dominated the cover and got most of the pages, was a young man, dressed in a blue suit, fedora, and small mask. To fill out the rest of the pages, two other minor characters were added: Mr. Mystic and Lady Luck. The latter was drawn by two other artists before Klaus Nordling took over the strip. Nordling had an unlikely background for a comic book artist; he was born in Finland in 1915 and later immigrated to the United States with his parents.

Nordling's *Lady Luck* was a young, attractive blonde socialite in a penthouse, who donned a green dress, hat, and veil that partially concealed her face, but not her emerald earrings. She responded to any serious crime with alacrity and quickly subdued gangsters and robbers with her brain and athletic prowess. *Lady Luck* was successful enough in 1943 to move from the Sunday supplements to a standard comic book, *Smash Comics.* Within six years, she was so much more popular than the other crime-fighters in *Smash Comics* that the title was changed to her name, *Lady Luck,* in December, 1949. But her newly named comic book lasted for only five issues and was discontinued in the summer of 1950.

But obviously someone at NBC remembered her because ten months later, she was virtually resurrected as *The Lady in Blue,* a young attractive blonde socialite in a penthouse who donned a blue dress, hat, and mask that partially concealed her face, but not her sapphire earrings. This juvenile adventure series debuted on May 5, 1951; it was a 15-minute Saturday morning program. Although it was a transcribed series, and ran until December 8, 1951, only two known copies, both from May 1951, have survived.

Even for a kid's radio serial, this program was simplistic in scope and illogical in story line. The central premise was set forth in the introduction to the first episode.

SOUND: Telephone ringing, and receiver being removed from cradle

LADY IN BLUE: The Lady in Blue

ANNOUNCER: The Lady in Blue is known to the world of crime as a feared and hated criminologist and is known to criminals by only that name, The Lady in Blue. A young woman, never seen except in a blue dress, wearing fabulous blue sapphires, and always a blue mask, covering the upper portion of her face. She has a maid in the person of Harriet Higgins, who is probably the only living soul on earth who knows the true identity of The Lady in Blue. She is now in her luxurious apartment on the topmost floor in a centrally located downtown office building, dressing for a social evening at a nearby country club, as her real, but unknown, self.

The second episode had a different opening for the announcer to read, although it was almost as long as the one used in the debut show:

ANNOUNCER: The Lady in Blue...Perhaps the most famous, and yet least unknown, woman in the field of criminal investigation is The Lady in Blue. She is a young woman, apparently very attractive, though no one can be sure of her exact appearances, as she always wears a small blue mask over the upper portion of her face. Her constant companion is her faithful Cockney maid, Harriet Higgins. Tonight The Lady in Blue is...

Despite the fact that her identity was supposed to be super secret, almost everyone in town (police, crime victims, cab drivers) knew her address and telephone number. Moreover, she didn't actually have a dual identity, like her comic book predecessor. Lady Luck was the alter ego of Brenda Banks, a beautiful heiress, but The Lady in Blue was only The Lady in Blue and apparently had no day-to-day dual identity in which to relax from her labors combating crime. In the comic books, this stunning crime-fighter had one of two male sidekicks, depending on who was drawing the strip. Her first was Peecolo, a large Latino chauffeur, who was later replaced, when Nordling took over, with a small and inept assistant, Count DeChange. Her radio mirror image had only one helper, a British spinster, Harriet Higgins, who served as maid and chauffeur.

While the radio performers were certainly better than the scripts they held in their hands, their identities have yet to be discovered. The same is true for the NBC production personnel for this series. The woman portraying The Lady in Blue gave her a perky, self-confident, and carefree air, in contrast to the maid, who was fearful, cautious and devoid of a sense of humor. The following is typical of their exchanges, and took place as the heroine changed into her crime-fighting blue costume:

LADY IN BLUE: Hurry now, give me my mask and get into your dark gray dress.

HIGGINS: I ain't expecting no callers, Miss.

LADY IN BLUE: Really ? (Chuckling) Now, Higgins, you know you're coming with me.

HIGGINS: Oh blimey, do I have to play detective again, Miss?

LADY IN BLUE: Naturally. You wouldn't have me go to The Bachelors' Club unchaperoned at this improper time of night, would you, Dr. Watson?

HIGGINS: I ain't no doctor, and me name's Harriet Higgins.

LADY IN BLUE: In spite of that handicap, you'll play Watson to my Sherlock Holmes. Now run along and change.

In addition to having a female companion, The Lady in Blue had another characteristic that set her apart from Lady Luck in the comics. The latter was not a detective; she was a crime-fighting heroine who battled spies and saboteurs in the World War II era and later bested gangsters, robbers, and escaped prisoners. The Lady in Blue was a private investigator who worked for specific clients, who had been victimized by the criminal element, although she preferred a disguise, similar to comic book heroines. Even her assistant, Higgins, wore a gray veil when they were on a case so she would not be recognized either.

The unknown scriptwriter produced some interesting, and confusing, occasions in the story lines. When a thug arrived at the apartment of The Lady in Blue, her maid pretended to faint and then grabbed his ankles so her mistress could push him to the floor. Before he could get up, they had him tied down. Later, both a strong box and a wall safe were emptied of their jewelry by unknown thieves, which the scriptwriter termed "robberies," apparently not knowing that such thefts are actually burglaries.

A frequent character in these adventures was Lt. Jerry Green, a local police officer, played by a man who sounded like somebody's grandfather, so Green may have been close to retirement. At one crime scene, he greeted The Lady in Blue as "The Sapphire Siren." Another, but less flattering, nickname was bestowed upon her by a hoodlum who belittled her as "Little Blue Riding Hood."

Perhaps the series got better after May 1951. Possibly the gaps in the overall scenario were filled in later and the adventures became more exciting and entertaining. NBC kept this series on the air until December 1951 so one can hope that improvements were made. But until the later scripts, or more audio copies, are discovered, we will really never know.

MARY VANCE

During the summer of 1941, NBC aired a new half-hour series, *Miss Pinkerton, Inc.,* which featured a famous Hollywood married couple, Joan Blondell and Dick Powell. The title was derived from a fictional character, created by one of the most popular mystery writers, Mary Roberts Rinehart. However, the relationship of Rinehart, Blondell, and Miss Pinkerton is a little more involved.

From a historical perspective, the original Pinkerton was Allan Pinkerton, a barrel maker and Scottish immigrant to the U.S. in 1842. In the Midwest, he chased cow thieves part time and, in 1849, he became the first detective hired by the City of Chicago. Pinkerton was savvy, courageous, and incorruptible, and one year later, he founded the first detective agency in America, Pinkerton & Company. His firm protected President Abraham Lincoln before there was a Secret Service and Pinkerton was the Chief of Intelligence for Union General McClellan during the Civil War. The popular term, "private eye," comes from the famous Pinkerton logo, an unblinking eye captioned "The Eye That Never Sleeps." For over a century, a member of the Pinkerton family ran the company. It still exists today (the name was changed to Pinkerton, Inc. in 1965) and

Joan Blondell and Dick Powell in *Model Wife* (1941),
married in real-life at the time. (*Laura Wagner*)

primarily furnishes security guards to private industry and operates ar-
mored trucks.

Mary Roberts Rinehart was born in 1876 in Pennsylvania and grew
up reading the dime novels of *Nick Carter* and *Deadwood Dick*. At age 16,
she had her first published story, a detective yarn called "Richards, The
Shadower." Although she would go on to be trained as a nurse, married a
doctor, and had three children, her writing was so successful that it resulted
in her own publishing company, run by her family. Her biographer, Jan
Cohen, pointed out: "Her rise from genteel poverty to wealthy tycoon, via
her writing, may qualify her as America's first female Horatio Alger."

By 1900, her novels, usually serialized first in popular magazines,
were well accepted. In addition to her novels and mystery stories, she
wrote plays; her most successful theatre collaboration was *The Bat*, first
staged in 1920. Among Rinehart's 60 mystery novels, fourteen were made
into motion pictures, the first one, a silent film in 1911. Although she
created three different female sleuths, Letitia "Tish" Carberry, Louise Bar-
ing, and Hilda Adams, only the last one, nicknamed "Miss Pinkerton,"
remained popular through the mid-1900s.

The majority of Rinehart's crime fiction fell into the category which whodunit fans call the "Had I But Known" mystery. This type usually involves a woman who hears strange sounds in an old house, is privy to sinister conversations, but fails to relay this to the local police, thus putting herself in more danger. She ultimately stumbles upon some clue, which solves the mystery.

Hilda Adams, a 38-year-old spinster, was a private nurse, Rinehart's original profession. This lady crime solver appeared in two stories and two novels, beginning in 1914. Adams was borrowed from her nursing duties and sent undercover by the head of a detective agency (who was later an Inspector in the Homicide Bureau), George Patton. Of course, Rinehart's choice of the name for Adam's supervisor was purely coincidental since General George Patton graduated from West Point in 1909, and in 1914, he was an unknown junior officer in the U.S. Army.

Adams, "a trained operative," was not actually a detective, or as she told her readers in one of her first person narratives, "My exact relation to the Bureau has never been defined; I have never claimed to be a detective." Patton nicknamed Adams "Miss Pinkerton" and loaned her a snub-nosed revolver for dangerous assignments, but the unmarried nurse usually left it at home, where she lived with her canary.

In 1932, Hilda Adams made her third appearance in the pages of a Rinehart mystery; this time her nickname was the title, *Miss Pinkerton.* Warner Brothers bought the film rights to this novel, and released the film the same year, with Joan Blondell playing Adams. Lloyd Bacon directed the 65-minute motion picture. George Brent portrayed her police supervisor; others in the cast were Elizabeth Patterson, Ruth Hall, and Allan "Rocky" Lane, who would go on to become a Republic Pictures cowboy star. This film followed the Rinehart formula of old house chills, a black-cloaked killer, a dark and stormy night, and hands clutching from the darkness. There was only one departure from the novel; unlike Adams in the printed page, who eschewed firearms, Blondell seemed very comfortable holding an automatic pistol, as she did in the publicity poster for the movie.

During her eight years under contract to Warner Brothers, Blondell maintained a frantic pace; she appeared in nearly 50 films in that short time. Did she remember her 1932 film, *Miss Pinkerton,* when she and hubby, Dick Powell, were formulating a female sleuth radio show, and suggest using the title again? Probably, although it hasn't actually been proven yet.

Blondell and Powell had followed different paths to Hollywood and their close association at Warner Brothers, where they did ten musical films together, led to their marriage in 1936. Richard E. Powell was born in 1904 in Arkansas and studied music, primarily voice, as a youth. He attended Little Rock College but by his mid-20s, his good looks and pleasant singing voice got him movie roles as a baby-faced crooner, frequently opposite Ruby Keeler. Powell and Blondell were first cast together in *Gold Diggers of 1933* and they became man and wife three years later, after both had shed their respective first spouses.

Joan Blondell, whose first name was Rose, was born in 1906 in New York City, the daughter of vaudevillian comedian, Eddie Blondell. She and her sister, Gloria, were on stage before they reached first grade and they stayed in show business for the rest of their lives. Like many vaudevillians, Joan never finished high school; she was working full time in a stock theatre company by the time she was 17. She did a little of everything the next few years. Joan finally hit it big in on Broadway 1929 in *Penny Arcade,* in which she appeared with up-and-coming James Cagney. Warner Brothers bought the film rights, kept Blondell and Cagney, and released it in 1930 under the title *Sinner's Holiday.* It was a box office success and it launched the Hollywood careers of both Blondell and Cagney.

In 1941 neither Powell nor Blondell had much radio experience, although she had the principal role in *I Want a Divorce* on Mutual. But both of them sounded like microphone pros on *Miss Pinkerton, Inc.* J. Donald Wilson, who created the premise of the program, was its producer/director. The West Coast supporting cast he selected was exceptional; most of them were, or would be, leads in other radio shows: Hanley Stafford, Gale Gordon, Sara Berner, Elliott Lewis, and Ed Max. Stafford was famous as the bombastic father of *Baby Snooks* while Gordon was playing leading roles on *Fibber McGee & Molly, Our Miss Brooks,* and *My Favorite Husband.* Lewis and Max would later costar in *Voyage of the Scarlet Queen* and Berner, after many appearances on the *Jack Benny Show,* got the title lead in the comedy adventures of a lady sleuth in *Sara's Private Caper.*

Only one audio copy of *Miss Pinkerton, Inc.* has survived; it was the debut episode of July 12, 1941 which introduced all the characters. The show began when Mary Vance (played by Blondell), a law student at Cornell University, is informed that she has inherited the Vance Detective Agency in New York City, formerly owned by her Uncle Mike. The surname of Vance, while not nearly as famous as Pinkerton in detective fic-

Joan Blondell. (*Laura Wagner*)

tional lore, nevertheless was well represented. Mystery writer Frederic A. Kummer in 1924 created Elinor Vance, a feminine sleuth. Philo Vance, the detective, first appeared in 1926, originated by Willard Huntington Wright, under his pen-name of S.S. Van Dine. Also, the author of *The Lone Wolf* crime mysteries was Louis Joseph Vance.

After receiving the news of her inheritance, Mary takes the train from Ithaca to Manhattan and during the trip, a young man named Dennis

Michael Murray (Dick Powell) flirts with her. It is to no avail and she brushes him off, claiming her name was Ezmarelda Higgins. Her original intent, upon reaching New York City, is to sell the investigative agency and return to law school. When she learns that her agency has no female operative to guard the Bentley Emerald that night, she takes the job herself. A lavish formal social at the Bentley Mansion attracts a large crowd, including two famous jewel thieves, and the young man Vance had met on the train.

She is surprised to discover he was a New York City police detective on the "Society Detail," and this policeman is a little annoyed to determine her true identify and the reason she is at the Bentley Mansion. He issues this warning to her:

> MURRAY: Well, from mouthpiece to gumshoe in one day! Just go back to your law books. Crime isn't romantic; it's ugly, sordid, and it's rotten. You get mixed up with some pretty bad guys and your pretty head wouldn't be worth a nickel.

She, of course, ignores his advice and he dubs her "Miss Pinkerton." Later that evening, the jewel thieves shut off the lights, but Mary grabs the Bentley Emerald before they can, and she flees the mansion grounds in her automobile. When they capture her shortly thereafter, she convinces them she wants to join their criminal enterprise, so they take her and the emerald to their boss, a master criminal fronting as a jewelry store owner. Unlike his inattentive henchmen, the boss knows she is a detective, since her photograph had appeared in the newspaper that day. Before he can harm her, Murray and his police associates bust in, save Mary, and arrest the guilty trio. The first episode concludes in Mary's new office, with Murray bragging to her employees about his recent success. She confronts him in strong fashion:

> MARY: I had intended to sell it (the detective agency) but you men who think all women are helpless nitwits give me twelve kinds of a pain. Not only *am* I going to keep the agency, but I'm going to show you that I can out sleuth you in every direction.

If the plot seemed more of a comedy than a crime drama, it was supposed to be. Neither Blondell, Powell, nor the creator, J. Donald Wilson, had in mind anything more than an engaging, funny series that incidentally involved crime. The two writers on the series were Charles R. Miriam and Carl Foreman; the latter also wrote for *The Eddie Cantor Show*. The friendly rivalry of Mary and Dennis continued in *Miss Pinkerton, Inc.* through the summer of 1941.

About 1944 a young blonde on the set of *Meet the People*, Ella Geisman, whom MGM had renamed June Allyson, caught the roving eye of Powell, and their resultant affair ended his marriage to Blondell . Later Powell married Allyson, and shortly thereafter, Mike Todd, millionaire producer, became the third husband of Blondell. Powell's career continued successfully, as a tough guy and leading man in the movies, as a radio detective in *Richard Diamond,* and as a television actor and producer. When he died in January 1963, he was only 58 years old.

Blondell's career was almost as impressive as Powell's after their divorce. She would receive an Academy Award nomination for Best Supporting Actress in 1951 for *The Blue Veil.* When movie roles became scarce for her, she returned to the stage and also did a lot of television work. Blondell appeared in over 20 different television series, including three in which she was the co-lead: *The Real McCoy's, Here Come the Brides,* and *Banyon.* A lady of a very strong work ethic, she had appeared in over 100 movies in her lifetime; one of her last films was 1978's *Grease,* completed one year before her death. She passed away on Christmas Day, 1979, a victim of leukemia.

HELEN HOLDEN

The creators of radio's soap operas loved to repeat the first letter of each name in a show's title. *Hilltop House, Lora Lawton, Houseboat Hannah, Manhattan Mother,* and *Helping Hand* were a few examples of this tendency. But the top prize for alliteration must surely go to *Helen Holden, Government Girl.*

This series, a women's daily drama about the adventures of a young G-Woman in the nation's capital, debuted on the Mutual Network in March 1941 and aired six times a week until it went off the air one year later. As with most soap operas, it was a 15-minute program. There is no available data indicating its sponsor, and the presumption would be it was a sustaining program.

Despite the fact that over 300 episodes were broadcast by Mutual, not one single audio copy has yet surfaced, nor have any of the original scripts come to light. Radio historian John Dunning summarized *Helen Holden, Government Girl* as:

> "A romantic crime serial (which featured) Nancy Ordway as Helen Holden, young, single G-woman…Nell Flemming as Mary Holden, her aunt (and) Robert Pollard as David, her boyfriend."

Another respected authority, Jim Cox, in his book, *Radio Crime Fighters*, added the following description:

> "In one of the first daytime radio serials dealing with war themes…Holden was sworn to protect the homeland against enemy infiltration and aggression and faithfully carried out her tasks. Her boyfriend, David, and an aunt, print news journalist Mary Holden were on hand to add a domestic touch to the melodrama.

But the absence of factual data about this lady crime-fighter was lamented by Cox, who explained:

> "Such lapses exist when historiographers are unable to acquire or substantiate records, and particularly when no known tapes of a drama exist to enhance contemporary appreciation for a series that may have aired decades earlier."

The Mutual Network had few name stars in its soap opera roster, but all three leads in *Helen Holden, Government Girl* must have been broadcast minor-leaguers. Ordway, Flemming, and Pollard are only mentioned once in all standard vintage radio encyclopedias: the single entry of their Helen Holden affiliation. There is even some confusion about the spelling of the name of the performer in the title lead. Was it Ordway or Ondway?

In its 1941 annual yearbook, WDEF, Mutual's affiliated station in Chattanooga, Tennessee, devoted virtually all of its 32 pages to background information and photographs of the local station personnel and the programs originating at WDEF. But it contained occasional entries publicizing network programs, including a full page entitled "Drama." This page is filled with four large photographs of radio personalities who currently had the lead in a Mutual series: Joan Blondell of *I Want A Divorce*,

Nancy Ondway was *Helen Holden, Government Girl*

Keenan Wynn of *The Amazing Mr. Smith*, Linda Watkins of *We Are Always Young*, and Nancy Ondway of *Helen Holden, Government Girl*.

Of course, Blondell and Wynn were well-known Hollywood stars, who dabbled in radio between films. Watkins and Ondway must have been delighted to share this page with two motion picture luminaries, since at the time, neither Watkins nor Ondway had made any impression on the public. While it was an uncommon practice to cast relative unknowns in the leads of network shows, Mutual occasionally took such chances. After their respective series ended, neither woman had major success in radio, according to available evidence. Ondway apparently never had any significant network roles. Watkins, however, was in the supporting cast of two detective shows: *The Fat Man* in the late 1940s and *The Big Guy* in 1950, a cloyingly silly series. She was also in *Amanda of Honeymoon Hill* and thrillers, i.e. *The Chase*.

Just what Helen Holden actually accomplished in her year on the air as a Government Girl is still unknown. The cases she worked, the mysteries she solved, and the evil doers she may have put behind bars remain a puzzle for vintage radio researchers to discover.

MARY SULLIVAN

Policewoman was the only radio series about a lady crime-fighter which was based on an actual woman. Both the real Mary Sullivan, and her radio impersonator, Betty Garde, appeared on the program.

Phillips H. Lord, one of the more well known producers and directors in broadcasting's Golden Age, had concocted what he thought was a surefire recipe for radio success. He would create a new series, starting with the exploits of a famous person, and then hire that same person to narrate the radio program. Lord was never deterred by the fact that these celebrities seldom had any ability to perform in front of a radio microphone. After each star flopped, Lord kept his series going by replacing the famous person with a relatively unknown, but talented, radio actor.

In 1935 Lord created *Gang Busters* and hired a very well known law enforcement officer of that era, Col. Norman Schwarzkopf of the New Jersey State Police, who had led his region's investigation of the Lindbergh kidnapping. While an earnest fellow, Schwarzkopf sounded terrible on radio and soon Lord had him replaced with radio actors who pretended to be law enforcement representatives. Then, in 1939, Lord began a radio show to cash in on the popularity of flying, which he titled *Sky Blazers.* To narrate this show, Lord put Col. Roscoe Turner under contract. Turner was one of the most famous aviators in the world and had recently won several air races in the U.S. and Europe. But his talent did not transfer to radio so soon after the debut of this series, Turner was yanked, and a radio actor, with the designation of "Flight Commander" became the new narrator.

Lord, in 1946, located another public figure, who while not as famous as Schwarzkopf or Turner, was still a person of some renown. Mary Sullivan, a 63-year-old grandmother, had recently retired from the New York City Police Department, having logged 35 years of honorable service in that agency. A remarkable accomplishment, especially for a woman, Lord must have thought, and he bought the rights to her life story and signed her up to narrate her adventures on the radio in a series Lord called *Policewoman.* Betty Garde was chosen to impersonate Sullivan in the dramatic portion of this 15-minute weekly program which debuted on ABC on May 6, 1946.

Of course, the real Sullivan was a poor performer at the microphone; her voice was hesitant, flat, and she stumbled over her lines. In short, she was boring. Lord, accordingly, reduced her air time on *Policewoman* to as little as possible. But he kept her on the payroll for the entire run, probably just for that little dash of authenticity at the end of each episode.

Karl Swenson with *Policewoman* Betty Garde

The real Mary Sullivan was born in 1883, was widowed in Manhattan shortly after the birth of her first child. Her three brothers, all New York City cops, found her a job on the force as a police matron so she could support herself. It must have been difficult for a woman, even though she was Irish and had influential relatives in that agency, to advance administratively, but she did. Sullivan searched and interrogated female felons, did some undercover work, developed criminal informants, and rose gradually in the ranks. By the time she was ready to retire after 35 years service and several commendations, Sullivan was the Director of a separate policewoman's bureau within the New York Police Department.

Betty Garde, who played Sullivan on the air, was born in 1905 and raised in Philadelphia . She went from high school right unto the stage of a local stock company. Because she was tall and her voice sounded much older than her actual age, she was usually cast as an elderly woman. Garde made her Broadway debut in 1925, and although she continued to get theatre roles, by 1930 she had switched to radio acting, which provided more steady income than the stage. Her "older woman" voice was just as noticeable at the microphone so she was hired for spinster aunts and grand-mother roles long before she reached middle-age.

Whether on a soap opera, an anthology series, or a crime drama, Betty was almost always given the role of the oldest female character in the show. She was the mother in four series (*My Son and I*, *Joe and Mabel*, *Maudie's Diary*, and *We, the Abbotts*), had the title lead in *Mrs. Wiggs of the Cabbage Patch*, and was Belle, the wife of *Lorenzo Jones*. In addition, Garde was in the supporting cast of *The Fat Man*, *Morey Amsterdam Show*, and *The Big Story*. She returned to the stage when a good role was offered to her; in the original Broadway production of *Oklahoma* in 1943, she played Aunt Eller, the oldest woman character in that musical.

Although she was 39 years old when she got the title role in *Policewoman*, the fact that her vocal quality made her sound older was probably a major factor in choosing her for the part. In 1946, the real Sullivan was 63 years old, but, of course, no mention was ever made on the program of the age of either Sullivan or Garde. Nor did Garde play the role of Sullivan as other than a middle-aged woman.

Policewoman ran as a weekly show, on Monday nights for the first two months, and then on Sunday evenings for the next year. Each program contained assurances from the announcer that the weekly case was "based on an episode in the career of Lt. Mary Sullivan" but a generous poetic license was granted to the scriptwriters of this crime drama. Not only were most of the real Sullivan's actual cases fairly mundane, but many had occurred 20 or 30 years prior. *Policewoman* was not a period piece; all the radio adventures took place in the post-World War II era, long after the real Sullivan had retired from police work.

As with most of the radio productions of Phillips H. Lord, *Policewoman* had an ear-catching opening that was both simple and effective:

ANNOUNCER: Carter's Pills present:

SOUND: Blast on a police whistle

ANNOUNCER: (*On filter microphone*) Policewoman!

SOUND: Organ sting

ANNOUNCER: Tonight: The Case of the Scheming Bride-groom, based on an episode in the career of Lt. Mary Sullivan, for 35 years on the New York City Police Department!

SOUND: Organ sting

In these modernized adventures, radio's Mary Sullivan was diligent, resourceful, utilized up to date law enforcement forensic science, and seldom needed much help from her male counterparts in the department. On August 28, 1994, National Public Radio's *Weekend Edition* featured a segment by Harriet Baskas on radio's lady detectives. Baskas played the following excerpt from one of the three surviving audio episodes:

POLICE OFFICER: There's no question but that girl is the killer who murdered Mrs. Prescott. Mary, you've done a wonderful job.

SULLIVAN: Almost as good as a man could have done?

POLICE OFFICER: Almost, yes, if…(laughs) OK, Mary, you win. A better job. A *lot* better job.

Garde was backed up by a strong supporting cast, usually playing cops or robbers: Mandel Kramer, Frances Chaney, Walter Vaughn, and Grace Keddy. The organ music, for stings and musical scene bridges, was the work of Jesse Crawford. Sound effects on this series were also well done, but the identify of the personnel responsible has not yet been determined. Since each quarter-hour episode was complete from the report of the crime to its solution, the scripts were crisp and well-plotted.

The last episode, June 29, 1947, is one of the three surviving copies and in it, Sullivan almost single-handedly unmasked a clever killer, who after meeting his mail order bride, murdered her and absconded with her life savings. Sullivan answered his next advertisement for a wife-to-be, and passed herself off as a naive, uneducated woman of means. To lead him into a trap where her police colleagues are ready with the handcuffs, Lt. Sullivan utilized insightful interrogation, thorough review of records, fingerprint examination, and skillful planning. She proved that Uncle Shorty was guilty of the crime, not his nephew Pete, who was just an unwitting pawn in the deadly scheme.

Since this radio series was well-scripted, had superb performers at the microphone, and possessed excellent production standards, the only disappointing aspects for its listeners were the commercials and the sign-off by the real Sullivan. The value of Carter's Pills (sometimes called Carter's Little Liver Pills) was touted on every program. While there may be a way to praise a laxative on radio without offending a lot of listeners, these

1946-47 commercials had not quite achieved that level. Apparently, there were only a finite number of variations on the following:

ANNOUNCER: …and Carter's Pills will safely unblock your lower intestinal track and provide gentle relief from irregularity so that…

Sullivan's closing remarks were neither very entertaining nor interesting as the poor soul mumbled through her scripted message in a monotone. The below was her contribution on the final episode:

SULLIVAN: Pete Millford was (*pause*) found to be the, uh, innocent dupe of his uncle and released. Uncle Shorty (*pause*) was executed (*cough*) for the vicious murder he committed. This is the last program of our current series and (*pause*) I wanna thank my many listeners from coast to coast for their kind letters of interest, uh, and I hope they've enjoyed the show as much as I have.

When dramatic radio disappeared, Garde switched to acting in the movies (she was in *Call Northside 777*) and had many television appearances, including regular, major roles on *Edge of Night* and *The World of Mr. Sweeney.* She died on Christmas Day, 1989, at the age of 84.

In June 1950, CBS attempted to launch a similar series, *Police Woman-USA,* but it only got as far as the audition disk. Richard Sanville was its producer and director; he had experience in those same jobs on *Escape, Box 13,* and *Columbia Workshop.* He planned to run a series of episodes highlighting the accomplishments of women police officers, two of which were his researchers: Virginia Kellogg and Mary Ross. His cast is not identified in the audition disk, "The Red Rose Murderer," not even the woman who played the lead, Officer Sylvia Rollins, who eventually identified and arrested the crazed killer. The most recognizable voice in the audition disk is Parley Baer, who portrayed Rollins' supervisor, Captain Gray of Homicide. He, of course, began his unforgettable role as "Chester" on radio's *Gunsmoke* two years later.

The Better Halves

It's likely that most of the radio detective programs featuring a husband and wife were derived from the movies of Nick and Nora Charles, rather than their sole appearance in a 1934 novel, *The Thin Man.* In 1941 no less than four crime-solving married couples debuted on network radio: *Mr. and Mrs. North, Front Page Farrell, The Adventures of the Thin Man,* and *Michael and Kitty.* These were followed by three other couples after 1944, the last of which, Greg and Gail Collins of *It's a Crime, Mr. Collins,* arrived about 1955 as a syndicated program. All seven of these couples were well-to-do, sophisticated, affectionate to each other, and with one exception, childless. In reviewing their respective histories on the printed page, the silver screen, and network radio, it will become apparent which series copied another.

Nora Charles

As virtually every mystery fan knows, the husband and wife sleuthing duo of Nick and Nora Charles debuted in Dashiell Hammett's 1934 novel, *The Thin Man.* Although Mr. and Mrs. Charles appeared in only one novel, it launched them into a series of popular movies, radio programs, and a television series. Their successful transition was not unique for Hammett's creations; his character of Sam Spade, after his sole appearance in the novel, *The Maltese Falcon,* went on to equal fame in other media. However, that author's third major fictional character, the Continental Op, has been largely forgotten.

Hammett was cleverly described by Daniel Stashower in a recent *Smithsonian* magazine article as "A thin man who made memorable use of his Spade." The unlikely creator of Nick and Nora Charles was born Samuel Dashiell Hammett on a Maryland tobacco farm on May 27, 1894 and raised in Baltimore. He dropped out of school at age 13 to help support his family, and over the next few years he worked at a series of odd jobs, messenger boy, stevedore, yardman, moving to each new employment as his love for loafing was noticed by his boss.

At age 21, he joined the Pinkerton Detective Agency, where he acquired first hand knowledge of the sleuthing business, which he would later use in his crime novels. He then served a brief hitch in World War I as an ambulance driver, but never got overseas as he was diagnosed with tuberculosis. Then, as Hammett researcher, Stashower, summarized this period:

> "He (Hammett) spent the next couple of years checking in and out of Army hospitals, surviving on a small disability check. Far from a model patient, he drank, smoked, and chased women, even from his sickbed."

Later Hammett married one of his nurses, Josephine Dolan, sired a daughter, dropped his first name of Samuel, and became a self-taught fiction writer in desperation to support his small family, scribbling all night on the kitchen table. He wrote mostly for the pulp magazines, who paid by the word, and his first full length work, *Poisonville,* featuring a detective called the Continental Op (short for operative, which Pinkerton detectives were called then), was serialized in 1923 in *The Black Mask,* the premiere of the pulps.

In 1928, Alfred A. Knopf, a respected publisher, was sufficiently impressed with *Poisonville* to bring it out in hardback a year later under an improved title of *Red Harvest.* Its success spurred Hammett to write another novel with the Continental Op, *The Dain Curse,* and quickly followed it by his first, and only, Sam Spade novel, *The Maltese Falcon.* By now, his writing could support his dissolute lifestyle as a boozer and skirt-chaser. Then in 1930, this 36-year-old author met Lillian Hellman, a 26-year-old script-reader. Ignoring their respective spouses, she became his lifetime paramour and drinking buddy. Hellman would, of course, go on to become a playwright.

Despite the financial windfall of his book sales and the movie rights (the first of three *Maltese Falcon* films came out in 1931) Hammett's lavish spending always exceeded his income. His Hollywood writing contracts evaporated when the studios found out he was more interested in emptying whiskey bottles than he was in pounding a typewriter. So in

1932, strapped for cash, he lodged himself in the Sutton Club Hotel in Manhattan, and in six months, completed *The Thin Man.* Hellman knew how serious he was about this project since he occasionally went a full week without touching a drop of booze.

Among the many literary friends of Hammett and Hellman was humorist, S. J. Perelman, who had a dog named Asta, so Hammett bestowed that name on the terrier of Nick and Nora Charles. Hammett not only dedicated *The Thin Man* to Hellman, he told her that he had based the character of Nora upon her. She must have been flattered, although Nora in the novel was a rich, jealous, alcoholic who was a trifle spoiled.

The Thin Man was serialized in *Redbook* magazine and then the hardcover version came out in January 1934. Its sales were enormous and it was favorably reviewed by prominent literary critics including Alexander Woollcott and Sinclair Lewis. The rumor that it was a "dirty book" provided more publicity and pushed retail sales to even larger numbers. The basis for this rumor was a suggestive passage which followed an incident in the novel when Nick had a scuffle with middle-aged Mimi Wynant, while trying to physically restrain her from beating her daughter. After the conflict was resolved, Nora began the following conversation:

"Tell me something, Nick. Tell me the truth; when you were wrestling with Mimi, didn't you have an erection?"

"Oh, a little."

She laughed and got up from the floor. "If you aren't a disgusting old lecher."

When some organized churches and staid book publishers castigated the book, the publisher, Knopf, took out large advertisements in the major dailies which said that 20,000 people hadn't bought the book because of the question on page 192. As might be expected, people flocked to book stores, glanced at the passage on page 192, and then raced with the book to the cash register.

Hammett probably created Nick Charles' life and lifestyle as the ones that he longed for in his sober moments. In creating Nick's fictional biography, Hammett had him finish military service, then become an ace detective, and finally marry Nora. One year after the happy nuptials, Nora's father died, leaving her millions in cash, plus a lumber yard and a railroad. Nick immediately retired so he could devote his time to managing his wife's business investments, during the time they were not going to parties, first night events, and lavish dinners, all of which involved liberal applications of alcohol.

With the phenomenal success of *The Thin Man*, and the subsequent sale of the movie rights, plus the expectation of future royalties on all his books, Hammett gave up writing professionally and never wrote anything of consequence after 1934. His increasing alcohol consumption was surely a major factor in his lethargy, and also as literary critic Jonathan Yardley pointed out in 1983, Hammett was not necessarily lazy, but he had no ambition and always preferred play to work.

Hammett was not above accepting obligations he had no intention of fulfilling. In early 1934, King Features Syndicate, hoping to cash in on the value of Hammett's name, gave him a one year contract, and a large advance, to write a new comic strip about a fictional secret agent, "X-9," for which Alex Raymond would do all the art work. But the written product was sparse, and he seldom met a deadline, with King Features stringers contributing the balance. Within a few months, the comic strip syndicate had their legal division terminate Hammett's contract.

Some observers have argued that Nick Charles, a suave detective, was entirely different from the rough-and-tumble Sam Spade. However, other than the fact that he was married, Nick wasn't much different from Spade; both were tough, unfazed by physical pain, cynical, pragmatic, and even shared a similar philosophy of dealing with crime. On one occasion, in response to a question from Nora on the premise of reasonable doubt as applied to being innocent until proven guilty, Nick retorted:

> "That's for juries, not detectives. You find the guy you think did the murder and you slam him in the can and let everybody know you think he's guilty and put his picture all over the newspapers, and the District Attorney builds up the best theory he can on what information you've got and meanwhile you pick up additional details here and there, and people who recognize his picture in the paper—as well as people who'd think he was innocent if you hadn't arrested him—come in and tell you things about him and presently you've got him sitting on the electric chair."

Metro-Goldwyn-Mayer bought the film rights to *The Thin Man*, ordering the director, W.W. Van Dyke, to shoot it as quickly and cheaply as possible. He complied and once the cameras began rolling, he was finished in 16 days. His casting of the leads would prove to be inspirational; America fell in love with Myrna Loy and William Powell, who as Nora and Nick, were certainly more appealing, charming, and funny than the married couple in the book.

Neither of these film stars had an easy road to fame. Myrna Williams, a native of Montana, began in show business as a dancer when still a teenager. After she got in motion pictures, a Hollywood mystery writer, Paul Cain, talked her into changing her name to Myrna Loy. Thereafter, she starting getting a lot of Asian roles, including playing the evil daughter of Fu Manchu.

Powell, after some success on Broadway, came to Hollywood and spent a long time playing villains in both silent films and early talkies. He finally got into the movie mainstream after portraying the hero detective of three Philo Vance pictures. Loy and Powell first appeared together in a major motion picture, *Manhattan Melodrama,* which starred Clark Gable, which had the distinction of being the last movie that John Dillinger saw before his fatal shootout with the FBI outside the Biograph Theater in Chicago, IL on July 22, 1934.

The Thin Man delighted MGM officials with its box office receipts and a sequel was quickly ordered, entitled *After the Thin Man* so all the public would know it was derived from the first film. Neither MGM, nor most theater patrons, were bothered by the fact that *The Thin Man* of the novel and the movie, was not Nick Charles; he was Clyde Wynant, a skinny inventor whose disappearance led to multiple murders. Nevertheless, Nick became "The Thin Man" in the sequel and the four other films that followed it.

Although there were several changes made to the story in adapting the novel for the screen, many pages of Hammett's dialogue survived, some word for word, into the screenplay. But Asta, who was a female Schnauzer in the novel, was played by a male wire-terrier, whose trainer called him Skippy. The biggest difference was the tone set by the movie director and his two stars. Unlike the pair in the novel, this Nick and Nora were zesty, congenial, delightful, and while they did not refuse a drink, their consumption of distilled spirits was far less than their literary predecessors. In the novel, the Charles' usually awakened with hangovers and guzzled their first Scotch and soda with breakfast; on one occasion, Nora had a cocktail before and after her morning shower.

The six *Thin Man* motion pictures, released from 1934 to 1947 not only produced large box office receipts, they also created a reservoir of good will, waiting to be tapped by the radio industry. Powell and Loy did play their characters twice on network radio, both on *Lux Radio Theatre*, doing *The Thin Man* in 1936 and *After the Thin Man* in 1940. The first Nick and Nora series on the air, *The Adventures of the Thin Man*, came to the airwaves in July 1941 over NBC as a half hour show with Les Damon and Claudia Morgan. Hammett, of course, had nothing to do with the writing on the radio shows, nor did he do the screenplays for any of the half-dozen films of *The Thin Man*.

Except for brief periods, the series would remain on the air, on different networks, for different sponsors, until its final episode in September 1950. From 1948 on, it was called *The New Adventures of the Thin Man*. Morgan, both talented and durable, played Nora for all of the nearly seven years the show was on the air, but the actor playing Nick changed several times. Les Damon portrayed him for the first two years and was replaced by Les Tremayne, who held the role for about a year. The latter was followed by David Gothard, who did the role for Mutual until it ended in January 1949. It would have one more short run, the summer of 1950, with Joseph Curtain as Nick.

It doesn't matter if these performers were actually coached to sound like Powell and Loy, but they did, whether intentionally or not. So far, only six episodes from the entire run have been located, and it is doubtful there are more surviving copies. Three of the six were rebroadcasts by Armed Forces Radio Service (AFRS) and they were hosted by "Sergeant X," whose voice resembled, and was probably that of, Howard Duff. He was, of course, also playing a Hammett character on the radio, *Sam Spade*.

From the beginning, *The Adventures of the Thin Man* emulated the funny, sexy movies, not the gritty novel. While there were still murders to be solved on each broadcast, they were the unbloody, Oh-My-Goodness variety. Asta, the household pet, did nothing in the novel, except infrequently bark and lick Nick's bare toes. In the movies, Asta got plenty of screen time, and even sniffed out clues that advanced the solution of the mystery. However on radio, Asta was usually written out of the script; Gilbert Mack, who specialized in animal sounds, was the voice of Asta , so Mack obtained little work on this series.

The series was billed as a "mystery-comedy" from its first year. However, in the early years, most episodes dealt more with adventure, with just a sprinkling of comedy. But the characteristic urbane mystery and sophisticated excitement disappeared from the scripts during World War II, so that by the late 1940s, all that was left was comedic silliness and sexual innuendo. Regarding the latter, radio programs, despite censorship, could get away with a lot more than the visual medium of film. Most of the radio shows began, and ended, with the couple in bed and Nora cooing in the ear of "Nickeee-darling." Claudia Morgan could make her voice so sensual that even her commercials aroused some listeners.

> NORA: If we can't sleep tonight, I just pour a cold glass of Pabst Blue Ribbon beer, and do this (…long suggestive silent pause…) and say, "Goodnight, Nickeee-darling."

The addition of Parker Fennelly to the cast in the mid-40s was an apparent attempt to raise the humor level. Fennelly, using his "Titus Moody" voice, portrayed Ebeneezer Williams, the Sheriff of Crabtree County. Since such a character would have no jurisdiction in New York City, the scriptwriters merely created situations where he would "drop in on his old friends, Nick and Nora." Sheriff Williams, or "Eb" as they called him, is in two of the surviving audio copies, and both times, he solved the mystery before Nick or Nora did.

In most of the early episodes, prior to World War II, long before the scriptwriters turned to fatuous plotting, Nick and Nora routinely solved murder mysteries together. As a pair, they found the corpse, examined the crime scene, questioned suspects, and arrived at the solution about the same time, although Nick usually beat her by a whisker.

On some cases, Nick sent Nora into a den of evil in an undercover capacity. In the episode dated October 10, 1943, called "Anniversary Necklace," Nora was dispatched by her hubby to the residence of a suspected killer, and she was under instructions to impersonate a Swedish maid and look for clues. Morgan must have had fun portraying Nora Charles with a fake Scandinavian accent in this adventure. Occasionally, Nora's sleuthing was briefly interrupted when she became the victim of a kidnapping, or getting lost, or being drugged, as she was in one story where a female associate slipped a controlled substance into Nora's drink, which rendered her semiconscious all night.

By the late '40s, the Charles' found themselves in what can best be described as Damon Runyon Land, where everyone had a funny name, humor was on every corner, and even crime was fun-and-games. Two of the surviving audio copies contain shows that fall into this category. In the one entitled,

"The Case of the Passionate Palooka," Nick and Nora were dragged into an inconsequential case of a purloined pooch, in which the main characters were named Atom Bomber Brickhead, Scoots Gillette, and Banana Nose Norbert. Nora filled the Charles apartment with stray dogs and put Nick in charge of their keep while she went out to interrogate the prime suspect.

The other show had an even nuttier plot, the July 15, 1948 program entitled "Haunted Hams," involved Nick and Nora traveling to a farmland region to witness the tribulations of a summer stock company, rehearsing in a barn. The country bumpkins who populate this adventure had outlandish names: Methuselah Marbrain, Belle Mae Begottis, Silas Salem, and Newt Newton; the latter claimed he got the job of Fire Chief solely because he was the only one who could play the tuba in the firemen's band. When Nora was offered a leading role in the company's play, Claudia Morgan could hardly conceal her delight in gushing the following for Nora's response:

NORA: If you think I ever had any ambitions to be an actress, being adored, thrilling audiences, having my name and picture in the papers, just being too, too wonderful, you're (…pause…) absolutely right!

Over the approximately seven years the series was on network radio, as it gradually evolved from a mystery-comedy to pure parody, a few aspects remained constant. As previously stated, many episodes began and ended with the merry married couple snuggling in bed. In addition to these program openings and closings, Nick and Nora usually played at least one other scene in their pajamas. A pseudo-sexual activity appeared in many episodes involving Nick taking a young woman over his knee and gleefully spanking her, for whatever reason the scriptwriters could create. Still another characteristic of the series was the presence of sexual compliments, by either Nick or Nora, i.e. "You're lovelier than sable underwear."

The Adventures of the Thin Man may have been taken off the air for the same reason that *The Adventures of Sam Spade* was terminated in 1950. Both Hammett and Hellman were longtime supporters of Communist groups and causes and they were eventually called to testify before the House Committee on Un-American Activities. Hammett was president of a front group which had put up bail money for the leadership of the Communist Party, who had been convicted of criminal conspiracy, and four of whom had skipped bail and disappeared.

Both Hammett and Hellman refused to provide any testimony to the House Committee; he was jailed for contempt, but no action was taken against her. In his later years, as were many entertainers, Hammett was cited for income tax evasion by the IRS. Hellman later claimed that the IRS attached all of his income from his books, movies, and television to satisfy his back taxes. These combined political and financial problems of Hammett caused radio executives to distance their networks from him.

Nick and Nora had one more chance at network broadcasting, in the fall of 1957 when their two seasons of NBC television began on Friday evenings, under the title of *The Thin Man.* The unlikely pair in the leading roles were Peter Lawford and Phyllis Kirk; the latter had been the television wife of Red Buttons two years prior. Asta, played by a fox terrier, was as skilled as his movie counterpart in locating clues and hogging screen time. A recurring character, created for the television series, was a gorgeous con-artist named "Blondie Collins" who was played by Nita Talbot. After the last episode of this television series aired on June 26, 1959, NBC broadcast reruns of the series during daytime hours for another season.

Hammett, a lifelong smoker and alcoholic, died of lung cancer in January 1961; he was 67 years old. Claudia Morgan, and her quartet of Nicks, continued in radio after *The Adventures of the Thin Man* ended and all five eventually did some television work. Morgan died in September 1974 at the age of 62.

PAMELA NORTH

If you sometimes confuse Jerry and Pam North with Nick and Nora Charles, you're certainly not alone. The two couples were so similar that even radio historian Vincent Terrace got them mixed up in his 1981 encyclopedia of the radio programs from 1930 to 1960 entitled *Radio's Golden Age.* Under Terrace's entry for *The Adventures of the Thin Man* we read:

> "Format: The story of the Norths, Nick, a private detective, and his beautiful, trouble-prone wife, Nora..."

In Dashiell Hammett's 1934 mystery novel, *The Thin Man,* Nick and Nora Charles were married, childless, wealthy, sexy, sophisticated, lived in a Manhattan apartment, loved parties, consumed a lot of alcohol, had late breakfasts, never cooked a meal, moved about in posh social circles, and solved murder mysteries. In the novels of Richard and Frances

Alice Frost of *Mr. & Mrs. North* (*The Stumpf-Ohmart Collection*)

Lockridge, Jerry and Pamela North were married, childless, wealthy, sexy, sophisticated, lived in a Manhattan apartment, loved parties, consumed a lot of alcohol, had late breakfasts, never cooked a meal, moved about in posh social circles, and solved murder mysteries. In fact, the only significant difference between the two couples was that the Charles' had a pet dog, while the Norths had cats.

The first novel featuring Pam and Jerry, *The Norths Meet Murder,* was published in 1940, although some of their adventures appeared earlier in *The New Yorker* magazine. The husband and wife writing team of the Lockridges was responsible for that mystery book and twenty-five more novels with the Norths; the last one was released in 1963, long after the radio and televisions versions had left the air. Despite the obvious imitation of the couple in *The Thin Man,* apparently neither Hammett nor his publisher, Alfred A. Knopf, took any legal action against the Lockridges or their publishers, J. B. Lippincott and Harper & Row.

Richard Lockridge did not limit himself to mystery books about the Norths. He was a very prolific author, and by himself, or with the collaboration of his wife, Frances, he produced dozens of other detective books. Many of these featured Captain Heimrich of the New York State Police or Lt. Nathan Shapiro. Occasionally Lockridge assigned them the same case, i.e. *Murder Can't Wait,* but usually they appeared in different novels.

The general dimensions and tone of the Norths was set forth in their first book. Gerald "Jerry" North worked for a Manhattan publishing firm, and his wife, Pamela, nearly always called Pam, was unemployed and filled her daytime with cocktails, talking to her cat(s), trips to the hairdressers, and gossip sharing. When one of their acquaintances was murdered, usually in chapter 2 or 3, the Norths discussed the case and Pam made some general inquiries, but the majority of the investigation was conducted by Lt. William "Bill" Weigand, a homicide detective with NYPD, who ultimately solved the case. Weigand, whom the Norths met in their first novel, remained a central character in the rest of their many novels, even when the cases strayed outside his jurisdiction. This trio of Pam, Jerry, and Bill, would become, as she described them, in 1958's *The Long Skeleton:*

> "We act like a mosaic. I mean, where does one leave off and the other begin? It's very nice, really."

Despite her claim, Weigand and the police do nearly all of the real work in solving each homicide and so entire chapters pass before the readers' eyes with no mention of the Norths, who were presumably attending another of their endless cocktail parties. But Weigand occasionally used the Norths, particularly Pam, as a sounding board for his theories on each mysterious killing.

Since neither of the Norths had any official law enforcement status, nor were they private detectives, the only way to obtain a corpse to set their novels in motion was to have a close friend, neighbor, or associate become a homicide victim. One would think that after two dozen murder mysteries, the Norths might run out of potential victims, but they did not. The Lockridges admitted, in the words of a character in one of their novels, that the Norths "are lightning rods for homicide." Moreover, when a local deputy sheriff in Key West, FL does not include the Norths in his investigative deliberations, Weigand telephoned him from New York and changed the deputy's view by describing the Norths as follows:

> "When they're around, things seem to turn up, useful things. Get them to help, if you can."

Obviously not all of the Norths' energy was expended in guiding the police, they also played bridge and tennis, went shopping, had dinner with friends at elegant restaurants, admired their cat(s), and consumed prodigious quantities of alcohol. In fact, one of the best ways of distinguishing one novel from another is to take a census of their cats or notice the drink favored by Pam and Jerry. In the first book, their cat is named "Pete" but is replaced by "Martini" for several books. During the 1950's novels, Martini was joined by Gin and Sherry. Finally, in their last book, the cats' names were no longer motivated by booze; they were called Stilt and Shadow.

As for the Norths' beverage of choice, in the early books they downed a large number of cocktails, primarily Tom Collins, martinis, and sometimes Bicardi Rum. Over the years, Pam gradually came to prefer a Bullshot or a creme de menthe, while Jerry usually drank gin mists (he accepted them with Beefeater's Gin only if the bar was out of House of Lords.)

It's almost surprising that the authorities got any assistance from the Norths in solving crime. Jerry was usually puzzled by each new homicide, and Pam, despite her supposed enthusiasm for the solution to a mystery, spent more time at cocktail parties, late suppers, and shopping than she did in investigative pursuits. Her few contributions to Weigand's efforts consisted of gossip heard at her hairdressers or relating the instincts of her cats; if one of her felines did not like a person, they were immediately suspect.

Regardless of the fact that the Norths were borderline alcoholics and they occasionally used their pets as lie-detectors, the popularity of their books remained high for over four decades. Their novels have never gone out of print, and in the 1980s, Pocket Book Library began reissuing all of the titles in paperback. Literary critic, Emma Lathen, described Pam and Jerry's appeal to mystery fans:

> "Like Conan Doyle's London, the Lockridges' New York has a lasting magic. There are taxis waiting at every corner, special little French restaurants, and perfect martinis. Even murder sparkles with big city sophistication. For everyone who remembers New York in the Forties and for everyone who wishes he did."

While the Norths may have not originally been as saucy as their counterparts, Nick and Nora Charles, by the late 1950s the Lockridges had begun to ratchet up the sexual texture of their novels. In the opening paragraph of 1953's *Curtain for a Jester,* the Norths were getting ready for a formal evening. Jerry, while adjusting his bow tie, was admiring Pam whom he could see in the mirror. In the next paragraph, Pam put on her bra, panties, and garter-belt. By the time the reader realized that Jerry had been gazing at a nude Pam, the Norths' conversation had turned to a recent "prank" where a two-way mirror had been installed in the shower room of a local college's women's dormitory.

Relatively few fictional crime-solvers (or their wives) made it to the Broadway stage, but the Norths did. A murder mystery drama, based upon the first novel, *The Norths Meet Murder,* was written by Owen Davis and it opened at the Belasco Theatre on January 12, 1941. It was produced and directed by Alfred De Liagre, Jr., while the set was created by one of New York's most honored set designers, Jo Mielziner. Peggy Conklin portrayed Pam North; she was very successful in film, on stage, and on radio. (See additional biographical details on her under Kitty Archer in Chapter 7.) The roles of Jerry North and Lt. Weigand were respectively played by Albert Hackett (who, with his wife Frances Goodrich, wrote screenplays such as *The Thin Man, After the Thin Man,* etc. for MGM) and Philip Ober; they were supported by seventeen others in this large cast. The Broadway production was a modest success; it ran for 163 performances and closed on May 31, 1941. Ironically, Conklin cut the audition disk for radio's *Mr. and Mrs. North* (with Carl Eastman) later that year, but when the network series was launched in 1942, two other performers got the leading roles.

In mid-1941, MGM , hoping to replicate the success of its *The Thin Man* series, released *Mr. and Mrs. North.* The screenplay was by S. K. Lauren, adapted from the stage-script of Owen Davis; Robert B. Sinclair directed this B movie, which ran just over one hour. The casting was unusual; Jerry was played by William Post, Jr., while Gracie Allen was in the role of his wife, Pam. The box office receipts were not impressive, probably because it was, in the words of film critic, Leonard Maltin:

> "...a comedy involving dead bodies being discovered; sometimes funny, sometimes forced, though it's interesting to see Gracie without George (Burns)."

Although neither the stageplay nor the motion picture version had achieved other than minimum success, NBC apparently thought that since the mystery novels were still selling well, a radio series with Pam and Jerry would be a welcome addition to their similar series, *The Adventures of the Thin Man.* Alice Frost and Joseph Curtin were chosen to play the title roles in *Mr. and Mrs. North,* which debuted December 30, 1942.

Curtin, a native of Cambridge, MA, was a child actor and had played juvenile roles in university stage productions at Harvard University. However, when he graduated from high school, he headed for Yale, which had offered him a better scholarship. Curtin dropped out of Yale after three years to pursue acting full time, doing summer stock in California and Shakespeare in the New York area. His first radio job was on the CBS series, *Roses and Drums,* in 1934, and soon his regular roles on several soap operas gradually drew him away from the stage. Curtin was 31 when he won the audition for the radio role of Jerry North.

Frost was one year older than Curtin; she was born in Minneapolis, Minnesota, the daughter of a Lutheran minister. She broke into radio playing comedy bits on *Stoopnagle and Budd,* but soon had recurring roles on several soap operas, plus anthologies including *Columbia Workshop, Town Hall Tonight,* and *Mercury Theatre of the Air.*

The radio series differed from the mystery novels of Pam and Jerry North in several respects. Their alcohol consumption and sophisticated sexiness disappeared almost entirely. The ever-present household cats, some of whom had sniffed out suspects, also were eliminated from the radio adventures. Those two police officials from the first book and the stageplay, Bill Weigand and Sgt. Mullins, survived but their appearances were greatly reduced to the point where they were totally absent from several episodes. It was obvious that the radio Norths would be more active in crime-solving so police assistance was not as necessary. To balance the departure of these various aspects of the Norths' adventures, pervasive humor slowly increased so that by the late 1940s, the series became more of a situation comedy than a mystery drama.

On the August 28, 1994 *Weekend Edition* of NPR, Harriet Baskas discussed the radio series with Alice Frost, both of whom concluded that Pam had been an equal partner in unraveling mysteries with her husband, Jerry:

BASKAS: Alice Frost, whose ability to mimic popular actresses during the 1930s and 40s earned her the title of "The Girl of a Hundred Voices," played Pam North for years. She remembers that, although her character was a little loopy, she was always essential to the story.

FROST: Well, that was the whole thing, you know, and Mrs. North got in trouble, and somebody got her out of trouble. She was just marvelous, you know, and her husband would say, "Now listen, don't monkey around with that. They're very tough people," and so forth. And I'd say, "Well, no, I won't. I'm just going to see what I can find out about it."

The series remained on NBC from December 1942 through December 1946, as a weekly half hour show, usually sponsored by Woodbury soap. After NBC dropped it, CBS picked it up six months later, on July 1, 1947, under the new sponsorship of Colgate-Palmolive, but with the same two leads, Frost and Curtin. This version went on successfully for over seven years, ending in mid-November 1954; the 30-minute weekly version had been converted to a quarter hour show that aired five times a week in October 1954.

One way to measure the popularity of a detective series was how often its leads were invited to be "guest detectives" on Mutual's highly-rated game show, *Quick As a Flash*. It was a charade quiz with contestants guessing a dramatized event. The closing segment had well known radio detectives create a whodunit to be solved. The Norths appeared in this slot fifteen times between 1944 and 1950; the detective couple with the next most appearances were the Abbotts with only four.

Over thirty separate programs have survived from this long run, the earliest extant episode dating back to December 1942, and many of these audio copies are of Armed Forces Radio Service (AFRS) rebroadcasts. The good news is that we would not have these copies today had the AFRS not preserved some of their disks. The bad news is that, as it did with other radio series, AFRS chopped the beginnings and endings of each program, deleted the cast and crew credits, and substituted innocuous recorded music for all the commercials.

Although Jerry remained a publisher, and Pam's sole responsibility was being his wife, the two conducted themselves as though they were full time detectives. In one episode, a woman, in the process of requesting their aid to recover her stolen jewelry, called the Norths "the famous detectives" and Jerry and Pam accept the compliment without correcting the lady victim. During the war years, the series gradually moved away from adventures involving homicides toward ones of felony property theft, although an occasional corpse still appeared.

Another discernible change in the series was a marked shift from danger and mystery to comedy and mischief. Earlier programs involved the solution of fairly complicated murders, the solution of which resulted in physical risk to the Norths. In one episode, Pam was drugged unconscious so she could be used as an alibi by a female associate trying to conceal a murder she had committed. *Radio Mirror* magazine in the mid-1940s had a regular feature in which they would obtain a typical script from a popular network show and then pose the radio actors in about six photographs to illustrate the story line. The Norths were highlighted in the July 1946 issue, in a story entitled "Murder for Two." Pam demonstrated her courage by squeezing into a dumbwaiter to trap the killer.

In a rare episode in which Jerry does not appear, "Pam Goes It Alone," his better half solved a homicide by herself and thus released a man from prison who had been wrongfully convicted of that killing. To do so, Pam went to three key witnesses and got them to recant their testimony by passing herself off as 1) an IRS Agent ("Would you like to see my credentials and continue this discussion in my office?") 2) a Police Woman ("Want to see my badge and come down to the station?") and 3) a newspaper reporter. The last witness penetrated her ruse by demanding to see her press card, but she got the information she wanted anyway. At the very end, she was assisted by Lt. Weigand, played by Frank Lovejoy.

By the late 1940s, the humor of the series had virtually smothered most of the action and mystery. Even the bad guys had funny lines. In a case entitled, "The Missing Sparkler," which took place during a rail trip by the Norths, when the purloined jewelry turned up missing from the compartment of the criminals who stole it, one of the crooks remarked, "Hey, there are thieves on this train." At the conclusion of this adventure, Pam discovered the stolen jewelry hidden in the collar of a dog owned by the evil-doers.

Most of the funny business derived from comedic situations and the zany conversations of Pam that followed. In an episode called "The Milk Caper," the Norths' car broke down on their sunrise return from a party and they obtained alternate transportation on a horse-drawn milk delivery wagon. A stolen ring in one milk bottle led them to a corpse, so Pam telephoned Sgt. Mullins of Homicide and this exchange took place:

MULLINS: Don't tell me you're going to start finding corpses at the break of day?

PAM: Well, we found this one, Mullins, and there may be more.

MULLINS: You mean there's one already? A bonafide murder?

PAM: As far as we're concerned, it is. We'll leave it to you to make it official. That's why I'm calling you. Jerry and I are going to take the diamond ring over to Mrs. Stewart's place but you'll find the body over on Hewlett Street, about six feet from Mrs. Breedon's milk bottle.

MULLINS: Whaaat?

PAM: You can't miss it, Mullins, it's homogenized. And the milk man will be waiting there with a gray horse.

MULLINS: We must have a bad connection. There's nothing but gibberish coming out this end.

There were other changes over the years. The Norths' theme song, *The Way You Look Tonight,* was replaced by the music of the Halo Shampoo commercial, at the request of the sponsor. As with most network shows that had a live orchestra before World War II, this had been reduced to a lone musician on a keyboard, to handle all the background themes and scene bridges. Alice Frost got a chance to show off her singing voice in a December 1947 episode, where she passed herself off as a vocalist at an underground club by claiming she was "Choo-Choo North" and then belted out her version of "Civilization."

In the mid-1950s, the humor and fun in the show began to recede. The Norths' adventures lost their comedic flair and so in a typical episode from that era, the only humor consisted of one closing quip by Pam or the preview of the next episode by the announcer at the end of the show. In one adventure, Pam suffered some minor injuries following a scuffle with a murder suspect so in the program's conclusion, she telephoned Jerry and told him to bring home three steaks. "We'll eat two and I'll wear one." At the end of another program, announcer Charles Stark reminded their listeners:

ANNOUNCER: Be sure to join us again next Tuesday when
the Norths are pigeonholed by a pigeon, chased by
a pair of love birds on the wing, and caught by a
murderer who has flown the coop.

For a brief period, *Mr. and Mrs. North* were on both network radio and television. With Frost and Curtin in their radio roles, being heard each Tuesday night at 8:30 PM, their fans could also watch Pam and Jerry on CBS-TV at 10 PM on Friday evenings, beginning October 3, 1952. Of course, the two shows had different casts; Richard Denning and Barbara Britton were the TV Norths and Francis De Sales portrayed Lt. Weigand. Denning, whose birth name was Louis Albert Heindrich Denninger, certainly had reason to pick a different professional name. He had early success in Hollywood and was in about 50 movies between 1937 and 1942 when he enlisted in the U.S. Navy. The *Mr. and Mrs. North* television series was not that popular and CBS yanked it, after one season, in September 1953. About 100 days later, NBC-TV brought it back to the small screen in January 1954 with the same leads, but this run was even shorter than the first; it was dropped on July 20, 1954.

A few months after, in October 1954, CBS attempted to increase listenership for radio's *Mr. and Mrs. North* by reformatting the show from a weekly half hour one into a 15-minute program five days a week with Denning and Britton at the microphone. This experiment failed and that radio series ended on November 19, 1954. But CBS was not done tinkering with it; ten days later the series was back on the air as a weekly half hour show. It must have been difficult for the long term team of Frost and Curtin to lose their jobs to Denning and Britton. One might speculate that the television division of CBS by this time had more power than the radio division, but whatever the reason, two veteran radio performers had to surrender their microphone to a younger pair from the former television program.

This new series, with Denning and Britton as Jerry and Pam, took a different tack with the tone and texture of the program, removing all of the light humor in favor of gritty, somber story lines of brutal crimes. In addition, the Norths found their roles reduced in each script so their supporting cast took over the bulk of the microphone time. In some episodes, the Norths did not even appear in the story for the first third of the program. The episode, "Death is Forever," began with an adulterous wife failing in her attempt to kill her lover, who had threatened to leave her for a younger woman. When the lover is rushed to the hospital in critical condition, the cheating wife begs her husband, the attending physician, to kill her lover on the operating table and make it look accidental.

Another episode from this era, "The Man with a Rifle," was even more brutal in plot and treatment. The story starts with an unknown sniper in New York City who viciously murders three men in three weeks, none of whom are known to the killer. When a ten-year-old girl accidentally witnesses the sniper during the third fatality, the psychopathic rifleman begins stalking the terrified little girl and her mother. With these types of stories, it was no surprise that this last version of *Mr. and Mrs. North* lasted only five months, and went off the airwaves on April 18, 1955.

Denning and Britton continued with their television and movie work for years after. He got the plum role of Governor Grey in *Hawaii Five-O*, which aired in primetime for 12 years (1968 to 1980), the longest run of a police drama in television history. Frost, although she never got a major role, stayed very busy in television and was on dozens of shows, including *Police Woman, Lassie, Gunsmoke, Hazel, Bonanza,* and *I Remember Mama*. She did some film work too, appearing in *The Prize* (1963) and *I'll Take Sweden* (1965). Curtin was less prominent in television but stayed fairly active until his death in April 1979 at the age of 68. Britton died a year later, in January 1980; she was only 59 years old. Frost and Denning both died in 1998 when she was 87 and he was 84. She died January 5th in Los Angeles; he passed away in October in Escondito, California.

JEAN ABBOTT

Jean Abbott was the creation of Frances Crane (1896-1981), an American mystery writer. This resourceful, fictional lady crime-solver was introduced in 1941 in Crane's *The Turquoise Shop* where Jean Holly first met Patrick Abbott, a private investigator from San Francisco, when he came into her shop in New Mexico. By the end of this murder mystery, they were in love. The following year, Crane released her second detective novel featuring this couple, *The Golden Box*, and by the time the last clue was explained, Pat had proposed marriage and Jean had happily accepted.

Later in 1942, the third novel with Jean and Pat, *The Yellow Violet*, reached the public. Although they were supposed to have their wedding performed at the beginning of this book, a complex murder and spy mystery absorbed all their time so that 200 pages later, the mystery was solved, but their marriage had still not taken place. They did, of course, eventually marry, and spent the rest of their wedded life unmasking murders, confounding extortionists, and bringing assorted thugs to justice.

From 1941 to 1965, the prolific pen of Frances Crane produced thirty murder mystery novels featuring her crime-solving Mr. and Mrs. Abbott. Her custom was to use a color in the title of these Abbott mysteries, such as *The Applegreen Cat* (1943), *Black Cypress* (1948) and *Murder in Blue Street* (1951). Although Pat's office was in San Francisco, Crane placed only about half of the Abbott mysteries in the Bay City. Jean and Pat were called upon occasionally to solve varied crimes in New York City, Dallas, New Orleans, as well as Hong Kong and other foreign locales.

Crane used a clever technique in writing her novels to insure that Jean was an integral part of all the action; the author made Jean the narrator of the mysteries. By writing in first person, through the words of Jean, Crane made her the primary focus of each adventure. Virtually nothing significant happened in any Abbott mystery unless Jean, the narrator, was present.

In the novels, Jean was bright, brave, observant, and skeptical; in one aside to the reader, she averred, "I doubt practically everything, until it's proved." She was a circumspect lady from tip to toe, but also she drank Scotch and smoked, in short, the perfect wife for a private detective. Pat, in his Bay office, located on the 9th floor of a skyscraper on Kearny Street, just off Market, appreciated his wife's wit and mental deductive powers: "Your intuitions certainly save a lot of wear and tear on my brain."

Jean was not the only tough and savvy woman in the Abbott mysteries. Most of the female characters created by Crane, whether good or evil, were unafraid and gutsy. Pat's secretary, Lulu Murphy, chased off a dangerous cab driver, bested a vicious bartender in a battle of wits, and packed a gun occasionally. In one novel, an elderly housekeeper pulled a gun on Pat. A European soprano stood up to Italian Fascist spies in another scene.

But it was Jean Abbott who demonstrated the most courage and calm valor while in harm's way. When a revolver was stuck in her face, Jean asked for a cigarette and then memorized the inscription on the cigarette case of the gunman. Another time, with two criminals battling Pat, the plucky lady smashed a teapot over one crook's head, grabbed the pistol he dropped, and then she shot the second one before he could stab Pat to death.

All of the Abbott mysteries were lively, logical, and likable. If Crane had any weakness as a detective fiction writer, it was her habit of providing lengthy and detailed descriptions of even minor characters. For example, the first time Jean met Pat's secretary, Lulu, she described her as follows:

> "(Lulu) was thirty-eight, pert, tidy, and Irish-looking. She had round, innocent gray eyes and short, wavy, black hair with plenty of gray in it. She wore a navy suit, a white poplin blouse, beige lisle stockings, and sensible black oxfords."

It would be reasonable to assume that when the Abbotts made their transition to network radio, this husband and wife team would have been substantially the same characters that they were on Crane's pages. But this was not what happened. Since Crane was only pocketing the royalties, while others wrote the scripts, Jean Abbott found herself demoted to the back seat.

Abbott Mysteries debuted in June 1945 on Mutual as the summer replacement for the popular quiz show, *Quick as a Flash,* which was sponsored by Helbros timepieces. It served well enough in that replacement slot to earn two similar runs in the summers of 1946 and 1947, each time starting in June and then going off the air around Labor Day, when *Quick as a Flash* returned.

Les Tremayne and Alice Reinheart, who were married in real-life, portrayed Pat and Jean the first two summers. Chuck Webster and Julie Stevens took over the roles for the summer of 1947. All four were seasoned performers with strong backgrounds in dramatic radio, soap operas, and romantic adventure programs. Radio historians have a difficult

Alice Reinheart of *Abbott Mysteries* (*The Stumpf-Ohmart Collection*)

time comparing the two couples, in their respective portrayals of Mr. and Mrs. Abbott, since not one audio copy has survived from any the programs aired by Mutual those three summers.

However, eight years later, in January 1955, NBC resurrected the series, gave it a new title of *The Adventures of the Abbotts*, and ran it as a Sunday evening sustainer. The introduction was as follows:

"The National Broadcasting Company presents *The Adventures of the Abbotts*, starring Claudia Morgan and Les Damon as Jean and Pat Abbott, those popular characters of detective fiction, created by Frances Crane. NBC invites you to join Pat and Jean each week at this time for another exciting recorded adventure of romance and crime."

Morgan, among her many successful radio roles, had been Nora Charles for nine years in the 1940's on *The Adventures of the Thin Man*, and Damon had played her husband, Nick, for about four of those years. The 1955 NBC scripts were by Howard Merrill, whose radio writing credits included *Sherlock Holmes* and *Secret Mission*. The co-producers were Ted Lloyd and Bernard L. Schubert. Lloyd had a network background of romance dramas, among them, *My True Story* and *Whispering Streets*. Schubert, on the other hand, had specialized in detective series; *The Falcon* and *Murder and Mr. Malone* were both his productions.

Only one audio copy from this NBC series has survived, a broadcast from the spring of 1955 entitled, "The Canary Yellow Sack." If this program was typical of its series run, it is apparent that it didn't live up to the NBC introduction of it as an exciting adventure of romance and crime. And unlike the calm and urbane Pat Abbott of the novels, Damon played him as tough, abrupt, and occasionally, short-tempered. Although he introduced Jean as "This is my wife; we often work on a case together," it's obvious that he neither wanted, nor needed, her help. Apparently the scriptwriter agreed; Jean was relegated to being merely a kidnap victim and spent most of the half-hour off-mike, waiting for Pat to rescue her.

The plot for this particular episode was a little silly, convoluted, and even illogical. It begins when Jean finds a classified advertisement which offers an old chromo-print of Zachary Taylor for $ 12.15. Subsequently it is learned that this advertisement, and others like it, are a secret code by which a local gang sets up meetings with their suppliers, i.e. Taylor Street at 12:15 am. This particular gang is in the shady business of selling black-market sleeping pills, without prescription, at flop houses and all-night theaters. Why they used such an elaborate, and nonspecific, system of communication, instead of a simple telephone call, was not explained.

The first murder takes place in Pat's office; a female client is shot from the hallway. After Pat finds no sign in the hall of the killer, he phones the police, and then dismisses Jean.

JEAN:	Where do we go from here?
PAT:	I know where you go—home!
JEAN:	Where?
PAT:	Home.
JEAN:	Now?
PAT:	Jean, how many times do I have to tell you that detecting is a very dangerous business for you.
JEAN:	Well, if you think that with that woman lying there murdered, with that mysterious code, that I'm going to—
PAT:	(Interrupting) I will not have you running around on these cases, Jean. I said: go home!
JEAN:	Well, of all the ridiculous—
PAT:	Jean!
JEAN:	(meekly) Yes, dear.
PAT:	How long has it been since anyone's spanked you?
JEAN:	(obediently) I'm going.

The above dialogue seems more appropriate to a stern father and young daughter, not a crime-busting husband and wife team. Nor does it match the persona of Pat and Jean that Frances Crane had so carefully crafted. Claudia Morgan, in the above exchange, probably wished she was still portraying Nora Charles on *The Adventures of the Thin Man*. For had her husband Nick raised the issue of spanking, Nora would have probably responded, "Oooh, Nickie, you're so mischievous!" However, it is clear to any listener of the above episode that Pat's veiled threat is a serious warning, not a frisky invitation to some kinky affection.

After Jean heads home, she is quickly kidnapped by the gang, who then tries to force Pat to stop his investigation of them or they will harm his wife. Later, Pat stows away in the back of the gang's truck, enroute to a warehouse hideout, where he locates his wife. Although they are discovered by the crooks while trying to escape, both are then rescued by the police who has just arrived at the warehouse. The local authorities had tracked the gang, in the tradition of Hansel and Gretel, by following a trail of sleeping pills that Pat had dropped on the highway from the speeding truck.

It seems likely that scripts like these were responsible for the series getting the ax after only six months. Frances Crane must have been relieved when it went off the air. She had probably forgotten all about it ten years later, when in 1965, she wrote her last Abbott mystery.

GAIL COLLINS

Most detective fiction fans are aware that *Mr. and Mrs. North* closely imitated Nick and Nora Charles of *The Thin Man*. However, a much stronger case can be made regarding *It's a Crime, Mr. Collins* copying the Pat and Jean Abbott mystery novels of Frances Crane.

It's a Crime, Mr. Collins arrived very late in the Golden Age of Radio; it was briefly a syndicated program and then Mutual put it on the airwaves in August 1956. However, all of the 24 audio copies in U.S. circulation are actually Australian productions from the same era. It was not uncommon in those days for U.S. radio programs to be replicated with Australian casts for broadcast in that country. In most cases, the same scripts were used.

In their first episode, Gail Collins introduces herself and her private eye husband to the radio audience:

> GAIL: I knew Greg was for me (when) he walked into my curio shop in New Mexico. Greg, tall, lean, slightly on the Western side. He's a private detective in San Francisco and he went to Mexico City on a secret assignment for the U.S. government.

Compared with Jean Abbott, Gail's story of how she met her husband, his description, and government duties, is virtually verbatim. In her 1942 mystery novel, *The Yellow Violet*, the third in the Abbott's series, author Frances Crane has Jean Abbott explain to the reader:

> "I was lucky to hook a tall, lean, interesting-looking Westerner like Patrick Abbott. We had met in my curio shop in Santa Maria, New Mexico. But he'd been yanked back to his San Francisco office to do some detecting for the U.S. government."

And the striking similarities don't stop there. Most of the cases worked by the Collins' have a color in the title: "The Blue Steel Fountain Pen," "White Plumes, Red Blood," and "The Fabulous Redhead." Most of the Abbott mystery novels by Crane also have a color in the title: *The Pink Umbrella, Black Cypress, Flying Red Horse,* and *13 White Tulips.* Pat Abbott, although his office was in the Bay City, traveled a great deal as his assignments took him and his wife around the globe to France, England and Mexico. Greg Collins, although his office was in the Bay City, traveled a great deal as his assignments took him and his wife around the globe to Italy, Argentina and Mexico.

Documentation is scarce on the cast and crews of both the syndicated and the Mutual versions; we know Mandel Kramer portrayed Greg Collins in the Mutual series, but historians have not yet determined who portrayed his wife, Gail. Kramer also played Lt. Tragg in radio's *Perry Mason,* and would later become the title lead in *Yours Truly, Johnny Dollar.* Kramer, who was born March 12, 1916 in Cleveland, Ohio, was successful in radio early in his career and was regularly heard on *Counterspy, The Falcon, Stella Dallas, Gangbusters,* and *Terry and the Pirates.* By the time he won the role of Greg Collins, he was already in the cast of the CBS daytime television series, *The Verdict is Yours.*

Regarding the Australian version, a radio researcher in Melbourne, Jamie Kelly, recently advised:

> "The series was recorded in Melbourne, Australia by Hector Crawford Productions in the mid-50s, and consisted of 52 half-hour episodes Greg Collins was played by Keith Eden, a well known Melbourne actor, and Gail was Mary Disney. The narrator, Jack Little, should sound familiar to American collectors as he came to Australia from the U.S. after World War II. He had been was involved with many AFRS productions as well as *Screen Guild Players,* etc."

The brusque style of the Keith Eden in *It's a Crime, Mr. Collins* was devoid of either affection or tenderness toward his wife, which never quite suited the tone of the series, or even the tag-lines of each episode:

ANNOUNCER: So remember, folks—

GREG; Where there is crime and romance, there you'll find Mister—

GAIL: And Mrs. Collins.

Although Mary Disney played Gail Collins in most episodes, we have not determined her occasional substitute, for example in "Red Hot Mama." While their voices were similar in tone, there was a slight contrast in the quality; Mary had a more playful voice, while her replacement sounded more cultured. Jack Little, the narrator, also doubled as "Uncle Jack," a character who appeared in every third or fourth episode. (Note: this Jack Little was not "Little Jack Little," the pint-sized U.S. singer, whose real name was Leonard Little.)

In the tradition of Jean Abbott, both in the novels and the radio series, Gail Collins was not only the wife and sleuthing partner, she also narrated the story. Each episode of *It's a Crime, Mr. Collins,* began with Mrs. Collins providing a variation of the below:

GAIL: My husband's a private detective. I'm Gail Collins
               and I'll be right back in a minute to set the stage
               for our puzzling crime.

At the commercial breaks, Gail returned to the microphone and encouraged the audience to stay tuned:

GAIL: Keep your ears pinned. I'll be back in a moment
               with the rest of our story.

In nearly every adventure, Gail found important items of evidence, uncovered possible motives, accurately foresaw potential danger, noticed vital clues Greg missed, and not infrequently, she solved the case. In "The Brown Bag" episode, she was the first to notice their Spanish-speaking cab driver in Mexico City was listening to an English language station and their cab was not headed toward the destination Greg had requested. Gail's inventory of the clues in "The Rockaby Murder" led to the solution of a mysterious stabbing death. Her deductive powers were so acute in "The Brown Alligator Briefcase," she engineered a ruse which identified the murderer, and she went to accuse him while Greg was on another lead. Confronted with Gail's evidence, the killer admitted his guilt but then attempted to silence her permanently by burning her alive.

Gail, akin to Jean Abbott, was no stranger to danger and she handled each crisis with valiant grit. The night Greg picked the lock on the window of the suspected killer's hideout in "The Rockaby Murder," he invited his wife to enter before him, saying "Ladies first" and so she did. During the adventure of "The Pink Lady," the couple was on the street when Greg was suddenly wounded by the gunfire of an unseen assailant; Gail calmly pulled him to safety and tended to his injuries.

The assignments of Greg Collins tended to parallel those of Pat Abbott since many of their investigations originated at the request of the U.S. government. In "The Brown Bag," Greg was on a secret mission in Mexico, and in "The Brown Alligator Briefcase," he had just concluded a classified job in Rome, Italy. But Greg and Gail did not have to depend on federal matters to take them abroad; in "The Yellow Chrome Death" an old friend invited them to his tobacco plantation near Buenos Aires.

Since radio historians have not yet determined the writer(s) of this series, it's not possible to praise, or affix blame, for the highs and lows of the various scripts. In "The Chrome Yellow Death," which took place in Argentina, a preposterous cause of death appeared when a bola (a gaucho device of three weights on leather cords, thrown to tangle the legs of a cow) was the murder weapon. Greg later explained, with a straight face, that one weight cracked the victim's skull while the other two wrapped around his neck, strangling him.

Nearly everyone, except the scriptwriter of "Red Hot Mama," would know that the islands of the Florida Keys are very flat and usually less than 40 feet about sea level. However, in this episode, Greg and Gail are traveling to an assignment on one of keys and come upon a mansion on the edge of a 300-foot ocean cliff. Another discordant note regularly occurred, not in the script, but in Greg's Australian pronunciation of a controlled substance which he repeatedly called "merry-wanna."

Some of the better moments in the programs come from either the scriptwriter's imaginative dialogue or the performer's interpretation. Gail, in an aside to the radio audience, in "Red Hot Mama," described a seductive woman stalking Greg:

GAIL: There she was, the high-powered sports model.
           When she saw Greg, she revved her motor a little.

At the beginning of "The Rockaby Murder," which took place in the desert town of Dry Gulch, whoever was playing Gail delivered her introduction to the episode in a whimsical Southwestern drawl which was quite delightful. Another time, in the middle of "The Pink Lady," Gail humorously teased her husband by saying:

GAIL: Sorry, Hawkshaw, we don't have much of a case yet.

It's a Crime, Mr. Collins on Mutual ended on February 28, 1957; it was the last husband and wife detective team in the U.S. when it went off the air. Mandel Kramer, who at that time was already appearing on some television shows, supplemented it with other radio work. He had the distinction of being the last title lead in *Yours Truly, Johnny Dollar*; he played the insurance investigator from June 1961 until it was finally canceled on September 30, 1962. This departure marked the end of an era since it was the last crime drama series on network radio. Kramer had a long-running role on the CBS daytime television series, *Edge of Night,* a carbon copy of *Perry Mason.* He was also in a few films, including *Fighting Back* (1982), a modern day vigilante movie. Kramer was 72 years old when a heart attack took his life in January 1989.

SALLY FARRELL

In 1993, former radio star Florence Williams recalled the day during the summer of 1942 when she got her first job on a daily network program. A young, struggling performer who worked at a variety of temporary jobs in Manhattan to supplement her meager radio earnings, Williams had just heard of some vacancies at a nearby munitions plant.

> "I was debating about leaving radio and taking a job in a local munitions factory; the work was steady and the pay was good. Just about that time, I was called in to audition by Frances von Bernard, one of the casting directors for Air Features, owned and operated by Frank and Anne Hummert. It was for *Front Page Farrell* on NBC. I was chosen to costar with Richard Widmark and our show was on five days a week. The Hummerts were very good to their actors so we got to play parts on their other shows too...How grateful I was for all of it! I stayed on *Front Page Farrell* to the end of it and it lasted over eleven years. Richard Widmark left after two years, so Billy Quinn took his place. Later, when Billy was drafted, Staats

Florence Williams, of *Front Page Farrell*, on another series (*The Stumpf-Ohmart Collection*)

Cotsworth became Farrell and he stayed in the role until we went off the air."

Some radio historians cite Virginia Dwyer as the voice of Sally Farrell, prior to Williams taking over the role. While no current radio references document William "Bill" Quinn, a very successful radio actor, as ever playing David Farrell, a few of those books list Carleton Young as "David" between Widmark and Cotsworth. Contradictions between performers' memories and citations from reference sources are not uncommon. But it seems unlikely that Williams would not accurately remember her three costars on her longest running daily network series.

The Hummerts, the most productive radio producers in the history of radio, were directly responsible for creating and maintaining at least 60 series, about half of which were successful soap operas. This married couple developed the shows, lined up the advertisers, roughed out plot outlines, and hired all the writers and performers. *Front Page Farrell* began on Mutual in June 1941 and ran on that network until March 1942, when it was dropped. (This is probably the version that Dwyer and Young starred in.) In September 1942, NBC resurrected the series, replacing the leads with Williams and Widmark. This series was sponsored by American Home Products Corporation, advertising a variety of their many products, chiefly Kolynos Toothpaste, Anacin, and Aerowax.

Front Page Farrell was one of about a half-dozen hybrid series that were part soap opera and part detective mystery, as were *Perry Mason* and *Kitty Keene, Inc.* But over the years, this series became more adventure oriented, until by 1950 the crime-fighting exploits of David and Sally Farrell replaced the domestic tribulations of its soap opera roots. It was about that time that the program abandoned the staple characteristic of the soap opera genre: the meandering plotline that took months to resolve a crime or a crisis. From 1951 until the series finally ended in March 1954, Mr. and Mrs. Farrell solved every case in just five daily episodes. As vintage radio historian Jim Cox described the revised format:

"No other show in daytime radio could boast the resolution of a crime a week. David Farrell was on a roll, aided and abetted by producers who decided to help him grab more and more of those page-one bylines."

Florence Williams grew up in New Hampshire and was acting in nearby regional stock companies by the time she was in high school. There was a regional theatre in Peterboro, NH where she was apprenticed, and nearby was the McDowell Theatre, where Thornton Wilder was in residence. He enjoyed her performance as the co-lead in *Romeo and Juliet* but warned her to get rid of her "white mice" (her sweet but high-pitched voice) or she would never make it in the acting profession. She took his advice, improved her voice and her acting skills, and was later accepted in the repertory company of The Goodman Theatre in Chicago.

After one season in Chicago , and a summer at the Cleveland Play-house, a 20-year-old Florence Williams headed for New York City and instant poverty. For two years, she barely made enough money for room and board, eking out her subsistence with a series of part-time jobs. She recalled a job at Macy's Department Store where she demonstrated a new antiseptic rubber doll for children called "Hi-Jean." Williams even picked up some crime prevention jobs; one restaurant hired her as a detective to catch patrons trying to sneak out the back door without paying.

Finally her big break on Broadway came, playing a 14-year-old girl in *Madchen in Uniform*. The critics loved her but hated the play and it closed within two weeks. However, the favorable exposure landed her in a series of major roles in plays which starred Judith Anderson, Lillian Gish, and Tallulah Bankhead. Despite her limited success, there were bleak periods between productions during her ten years as a stage performer, and finally her radio friends convinced her that network jobs were more stable than anything in the theatre. Her transition from the footlights to the microphone resulted in another financially lean year, but once she got the job on *Front Page Farrell*, her troubles were over.

Her costar, Richard Widmark, was born the day after Christmas, 1914, in Minnesota but grew up near Princeton, Illinois where his folks had relocated. After graduating as a theatre major at Lake Forest College in Illinois, he taught drama for two years until 1938. Widmark made his radio debut in *Aunt Jenny's True Life Stories*, and then he alternated between stage and radio work in Chicago. He moved to New York City in the early 1940s. Although many actors were being drafted in World War II, Widmark's perforated eardrum rendered him unfit for military service. From 1942 to 1946, he made a modest living, with roles on daytime radio, and appearing in occasional Broadway plays at night.

In early 1943, about six months after he took over the title role in *Front Page Farrell*, Widmark was cast as Lt. Lenny Archer in a Broadway drama, *Kiss and Tell*, which opened on March 17th. This play told the story of the Archer family, including Corliss, and her two chums, Dexter Franklin and Raymond Pringle. Vintage radio fans know that these characters were also on the juvenile comedy on CBS radio, *Meet Corliss Archer*. The radio series actually predated its Broadway relative by two months but the two were entirely different. The stage play's crisis story line focused upon the pregnancy of an older sister, who was supposedly unmarried, which was serious business in the mid-40s.

Kiss and Tell had a very long run, 956 performances, but Widmark did not profit from it. He left its cast in November 1943 to take a bigger and better role in another drama, *Get Away, Old Man*. (For you trivia buffs, the actor who replaced Widmark in *Kiss and Tell* was Kirk Douglas.) Widmark must have been quite disappointed when *Get Away, Old Man* closed after fourteen performances because it took him about a year to find another Broadway role, and of course, *Kiss and Tell* was still running during that time. About 1946, he relocated to Hollywood to start his movie career. One year later, he terrified film audiences as he laughingly pushed an old lady in a wheelchair to her death down the stairs and *Kiss of Death* (1947) resulted in his nomination for an Oscar for Best Supporting Actor. Widmark was on his way.

In *Front Page Farrell*, David was the star reporter for the mythical *New York Daily Eagle* and every potential news story led him and his wife to solve a mysterious crime. Each 15-minute episode began with the announcer summarizing the case thus far. (Since he was more detective than reporter, David called his adventures cases, not stories.) Here's a typical introduction, from August, 1949:

ANNOUNCER: For the second time, a dream of Ramona
Petrie's has borne out with terrible accuracy. First,
the professional tea leaf reader told David Farrell
that her occult powers had revealed to her the
location of some stolen high explosive detonators—
and a policeman who was sent to check was killed.
Soon after David began investigating this murder,
Ramona had another dream and she warned David
Farrell that if he left his home that night, he too
would be killed in an explosion. But a phone call
from an unknown man, who promised important
information, lured David and Sally out of their
apartment. The man failed to meet them, and as
the Farrells returned home, David's hand on a light
switch set off a violent explosion.

Undercover operations were almost a tradition for Sally and David. In the summer of 1948, working "The Case of the Fatal Smile," their investigation focused upon an unusual extortion racket in which a carnival roustabout instigated a scuffle with a wealthy young man. In the ensuing fight behind a tent, the carnival worker pretended that the young man had struck him with a fatal blow and feigned his own death. Later the extortioner contacted the parents of the young man and demanded money to conceal the supposed death or the son would be charged with murder. After two such extortion demands were paid, David and Sally, under the guise of drifters, joined the carnival and got jobs operating a shooting gallery under a tent.

Another case which involved Sally going undercover was illustrated in the October 1946 issue of *Radio Mirror* magazine, using posed photographs of Florence Williams and Staats Cotsworth. The case began with a mysterious shooting at a beauty shop which alerted the Farrells to illegal activity on the premises. Sally, in the guise of a wealthy patron, went to the shop and learned that Lizette, the owner, was blackmailing her customers after recording their conversations with each other. Later, David sent Sally back with bogus, but realistic, jewels to pay off the blackmailer. Lizette discovered Sally's true identity and threatened her with a bottle of acid when David crashed in and rescued his spouse.

Frequently an otherwise inconsequential incident soon developed into a series of felony crimes that the Farrells had to solve quickly. The December 1951 issue of *Radio TV Mirror* magazine provided a summary of David and Sally's crime-fighting that week:

> "FRONT PAGE FARRELL: A manicurist inherits a fortune and almost at once reporter David Farrell is involved in 'The Little Blue Hat Murder Case.' A battle, which begins when a hat-shop owner accuses the manicurist of being the wrongful heir, ends when the hat-shop owner is killed. David and Sally become entangled in a dope racket before they solve the murder."

Although approximately 2,900 episodes of this quarter-hour program aired during the period 1941 to 1954, only thirteen shows in audio form have survived to the present day. Listeners heard the last live program on March 26, 1954. Florence Williams eventually returned to New England and the regional stock companies that she loved. She was active in the footlights and backstage in several theaters, and in her 70s, did a one-woman show on the life of Emily Dickinson. Each year, from 1984 to 1994, she took part in the Friends of Old Time Radio Convention in Newark, New Jersey, and portrayed many characters, from babies to grandmothers in that convention's re-creations of vintage broadcasts. Williams was 82 years old when she died in March 1995. (For additional information on Staats Cotsworth, see the entry of Ann Williams in chapter 4.)

DEBBY SPENCER

A lively and appealing series, *Two On A Clue* was another one of those hybrid shows, part detective adventure and part soap opera, as were *Kitty Keene, Inc.* and *Perry Mason.* But this program had several unusual and entertaining qualities that placed it above the other two, including the crisp and humorous scripting of Louis Vittes (rhymes with "kitties"), the writer on this CBS daily 15-minute show, and the direction of Harry Ingram, who treated it like an evening mystery program.

Unlike the other Mr. and Mrs. crime-fighters on network radio, *Two On A Clue* featured a family: father, mother, and a youngster. The Spencer family consisted of Jeffrey, his wife, Deborah, and their juvenile son, Michael, who were respectively called, Jeff, Debbie, and Mickey. All three roles were filled by very competent radio performers.

Ned Wever, the voice of Jeff, was a native New Yorker, graduated from Princeton, and moved from small Broadway roles to a very successful radio career. He was in several crime adventure shows, usually as the good guy, including the title leads in *Dick Tracy* and *Bulldog Drummond.* Occasionally he played characters on the other side of the law; he was "The Wolf," who was the nemesis of *Superman,* and Wever also portrayed both thugs and cops on *Gangbusters* and *Mr. Keen, Tracer of Lost Persons.*

His charming wife was played by Louise Fitch, a redheaded and dimpled 29-year-old woman, all of which characteristics writer Louis Vittes also ascribed to Debby Spencer. Fitch, originally from Iowa, was a graduate of Creighton University, and like Wever, came to radio after working on the stage. She kept very busy in a variety of broadcast roles, including several soap operas (*Manhattan Mother, We Love and Learn, Road of Life,* and *Women in White.*) Fitch had a wide range; she was Astra the Slave Girl on *Light of the World,* Nancy on *That Brewster Boy* and Betty Lou on the comedy, *Mortimer Gooch.* Despite her success at the microphone, *Two On A Clue* was one of her few leading roles.

The third member of the family was nine-year-old Mickey, played by 14 year old Ronald Liss. Mickey had the same name and age of the son of scriptwriter Louis Vittes who utilized some of the mannerisms he observed in his own son in creating the character of the Spencer lad. Liss was a real veteran by the time he got this role since he had been acting on radio for almost 12 years. He was no stranger to crime-fighter shows since he also played Robin, when the Dynamic Duo appeared on *Superman*, and he helped chase lawbreakers as Scotty on *Mark Trail*.

Vittes, a New York native, was born in April 1911. He briefly attended Columbia but soon got into the radio business, finding his niche as a scriptwriter, specializing in detective mysteries. He was the primary writer on *Nero Wolfe*, *Affairs of Peter Salem*, *Mystery of the Week*, *The Saint*, and *The Lone Wolf* plus he was on the writing teams for *Adventures of the Thin Man*, *Barrie Craig*, and *Mr. and Mrs. North*.

Two on a Clue debuted on October 2, 1944 on CBS, the same day as a soap opera, *Rosemary*, did on NBC. However, by 1945, *Rosemary* was on CBS at 2:15 p.m., and as such, was the middle of a crime-fighter "sandwich" since CBS' *Two On A Clue* aired at 2:00 p.m. and *Perry Mason* followed *Rosemary* at 2:30 p.m. Despite their time slot proximity, their tenure was not the same. *Two on a Clue* lasted only 16 months while *Rosemary* and *Perry Mason* were on the air for eleven and twelve years, respectively . However, the short duration of this series is in no way indicative of either the quality or popularity of the show.

Tune In magazine (April 1945) estimated that four million listeners were avidly following the adventures of the Spencers every afternoon. Jeff Spencer was an attorney, but he was always finding mysteries to solve, by choice, or, as he confessed to his wife on an early program, "Some people like caviar, other people prefer blondes, and I like mysteries." His resourceful wife shared his passion for thwarting evil-doers and accompanied him in his quest for crime solutions. Their son, Mickey, occasionally got involved in their adventures, but he was more likely to be gulping down a glass of milk in the kitchen while enroute to school.

Each program opened with the announcer, Alice Yourman, telling of the benefits of one of the three General Foods products, each of which was mentioned somewhere in every episode. LaFrance Bleach and Satina Laundry Starch filled most of the commercial time allotted, with Postum a distant third. However, rest assured that every time Jeff and Debby were at the breakfast table, they commented on how delicious their Postum was that morning. It was very unusual for a woman to be the announcer on any radio program, even a soap opera, but Yourman did a fine job. She was busy playing other characters on other network shows, including comedy and variety shows, and on Saturday morning she was heard as the mother of *Archie Andrews*. Yourman was usually joined in the *Two On A Clue* commercials by a male actor and they frequently played very curious combinations. In one, they both adopted "Old Plantation" accents to extol the joys of LaFrance and Satina, while in another commercial, Yourman played a vine-swinging monkey while her male counterpart portrayed a jungle explorer.

Each case the Spencers encountered took them about 18 or 20 episodes to solve, before moving on to another mystery. Since this was a daily show, that would mean approximately one month to discover the corpse, solve the tangled mystery, and bring the guilty to justice. Louis Vittes, in composing these lengthy adventures for the Spencers to unravel, gave each case a clever and colorful title. Among the mysteries our intrepid couple solved over the years were: "Case of the Red-Headed Doll," "Case of the Twice-Killed Corpse," "Case of Misunderstood Monkey," and "Case of the Actor's Exit." Their shortest adventure was 7 episodes, "Case of the Missing Umbrella," while their last mystery was also their longest, 37 episodes, "Case of the Dying Day Nursery."

A regular character on the show, representing the local law, was Sergeant Cornelius Trumble, the voice of tall, bookish John Gibson. A loyal and dedicated policeman, Trumble was usually a half-step behind the Spencers in the solution of any criminal enterprise, but the couple enjoyed his company and assistance. While they joked with him often, it was not in the nature of the "put-down" of city law enforcement officers by other radio detectives, such as *Richard Diamond* and *Sam Spade*. Here's a typical conversation between the Spencers and Trumble in the October 3, 1944 episode, in which a stranger is murdered on the Spencer's front porch and they are filing a report at the police station:

TRUMBLE: And you never saw this stiff before?

DEBBY: Not before tonight.

TRUMBLE: And when you came home, he was waiting for you on the corner?

JEFF: I like "lurking."

TRUMBLE: All right, was he lurking?

DEBBY: He certainly was.

TRUMBLE: Then what happened?

DEBBY: He accosted me.

TRUMBLE: With one or two c's? Does anybody know?

JEFF: Two c's, Cornelius, and usually with a leer.

Vittes' writing, which was terse, vibrant, and colorful, made each episode an entertaining quarter-hour that moved at a lively pace. His well constructed scenes played realistically, with a nice mixture of humor and danger. An uncredited organist, whose style resembled Harold Turner, punctuated all bridges between scenes with expressive and vivacious melodies.

The relationship of Jeff and Debby was based on mutual respect, equality, and a genuine love of each other that they frequently expressed when alone. In another episode, also from October 1944, when returning to their home in the evening, the couple notice a man, who appears to be armed, watching their house from their frontyard. The following conversation ensues:

DEBBY: Darling, what are we going to do?

JEFF: Nab our gun-carrying friend, if we can.

DEBBY: Theoretically, I'm partial to the idea, but actually
            he'd spot us if we got near him.

JEFF: Yeah, I know, therefore—

DEBBY: I love your therefores.

JEFF: How about my where-as's?

DEBBY: I love them too...Darling, let's not get killed if we
            can help it.

In most of the adventure mysteries that the Spencers were thrust into, this happily married couple were a team. While Jeff usually tried to protect Debby from physical danger, i.e. telling her to lie down in a taxi as he tried to get the license number of a passing car driven by a gunman, his motivation seemed to be his love for her, not any reflection on a lack of courage or spunk on her part. Both the scripts and Fitch's performances underscored the bravery and initiative of Debby Spencer.

The ratings were still strong when the series ended in January 1946 and reasons why it was taken off the air are still murky. But this could be said of other popular shows canceled in that era. Competition for most time slots was fierce, and the choices of sponsors, capriciousness of the networks, and other vagaries of radio production, cannot necessarily be translated now into logical reasoning. Regretfully, only three audio copies of *Two On A Clue* have survived.

All four of the performers, who had portrayed the Spencer family and their policeman pal, Sergeant Trumble, worked successfully in network broadcasting, until the growing power of television in the early 1960s decimated the prominence of dramatic radio. Louise Fitch continued on in the entertainment industry, and in the early 1970s had a regular role as Nurse Bascomb on *Medical Center* on CBS-TV. She died in September 1996; she was eighty-one.

Vittes relocated to Hollywood about 1948 and moved from radio duties to scriptwriting for the film and television industry. He wrote the screenplays for several westerns and science fiction movies, including *Pawnee, I Married a Monster from Outer Space, Showdown at Boot Hill, The Rebel Set, The Oregon Trail* and *The Eyes of Annie Jones.* Vittes also scripted several television episodes of *Last of the Mohicans, Rawhide,* and *The Virginian.* From the mid-1950s, he suffered from cardiac problems but was able to keep working. He died of a heart attack, in April 1969, just three days after his 58th birthday, leaving a wife and three sons. All three lads pursued careers in the communication arts. The eldest, Michael (whom the character of Mickey Spencer was based upon) became a television editor, producer and writer; now 68, he is retired in the Los Angeles area. Lawrence, 58, a former television reviewer, is now the editor for *Senior Life USA* in California, and the youngest son, Elliot, 53, is a college professor directing liberal studies at a Florida university.

KITTY PIPER

Most of us would probably bet that a radio detective series entitled *Michael and Kitty* would involve two main characters solving crimes. But we would lose this wager. For on this named program, the mystery solving team consisted of the trio of Kitty Piper, her husband Michael, and their omnipresent cab driver, Doc. These three were as thick as Jack, Doc, and Reggie on radio's famous *I Love A Mystery.*

Michael and Kitty (which, for its last two months, was called *Michael Piper, Private Investigator*) did not have a long run. NBC Blue Network aired this half-hour program on a weekly basis from October 1941 to February 1942. Its sponsor was a soft drink that no one, except perhaps beverage hobbyists, would recall: Spur, a product of the Canada Dry Bottling Company. Listening to the commercials some sixty years later fails to produce a lone clue as to whether it was a cola, a fruit drink, or a variation of ginger ale. The announcer extolled this "sparkling refreshment" and referred to this "new, delicious drink" as "America's favorite flavor at its very best." This would appear to be somewhat of an exaggeration since Spur, distributed only regionally, disappeared during World War II and has seldom been heard of since.

The troika of radio performers propelling this crime-solver series were all successful network veterans. Elizabeth Reller, who portrayed Kitty, had a distinguished tenure in the soap operas. During one period she was the co-lead in *Betty and Bob*, opposite Don Ameche, and also the wife on *Young Dr. Malone*. Reller also had principal roles on *Portia Faces Life* and *Doc Barclay's Daughters*. She was no stranger to detective shows; she played the assistant to *The Amazing Mr. Smith*, Keenan Wynn's crime show.

John Gibson, who was 37 years old when he got the role of Doc, made a career of portraying the sidekick, frequently to a crime-solver. He was Red, the best friend of *Don Winslow of the Navy*, and Sgt. Trumbull, who assisted the Spencers on *Two On A Clue*. Gibson is probably most familiar to vintage radio fans as the bartender, Ethelbert, a third-wheel on *Casey, Crime Photographer*. His radio skills got him regular roles on the premiere network programs, including *Columbia Presents Corwin, Dimension X* and its sequel, *X Minus One*.

The voice of Michael Piper was that of Manhattan-born Santos Ortega , certainly one of the busiest radio actors on the East Coast. He was often cast in the leading role as a detective and his skillful characterizations enabled him to handle any vocal specialty. (For a complete listing of his detective roles, see entry in this book under Della Street in Chapter 6: Me and My Gal Friday.) Ortega played major roles in most of the popular soap operas, network anthologies, and many adventure series.

Each episode of *Michael and Kitty* began with a brief musical flourish, from a live orchestra that played the bridges throughout the show, and then Mr. Piper would say to the radio audience, " Hello there! This is Michael Piper, Private Investigator. The case we bring you tonight is called (Fill in name of story.) As you know, Canada Dry and their many well-known bottlers of Spur, have been making these weekly radio shows possible. Well now, let's see, the whole thing began…" and he would set the opening scene.

In each episode our trio of sleuths together encountered a mystery, perhaps a robbery, a murder, or an assault. They traveled together in Doc's taxi to conduct surveillances, interrogate suspects, and check often with the police. A regular in the program was Inspector Shield, who was usually working the same case as Kitty, Michael and Doc. The cabbie and Shield provided humorous diversion by trading clever insults. Doc was always critical of Shield's crime solution theories and Shield, in turn, deeply resented the taxi driver butting into police business. When Doc made a joke at Shield's expense , the police officer would retort by calling him "Fred Allen."

One of the biggest mysteries about this series is how our three crime-fighters supported themselves financially. Michael and Kitty needed no paying client to activate their investigations. A case could begin as a puzzling crime that Kitty or Doc happened upon, while others originated as requests for assistance from the police department. The absence of a fee never slowed these investigators. Nor did Doc get many fares in his taxi business, since virtually all his time was occupied in providing transportation to Michael and Kitty. Apparently, this was one cabbie who preferred companionship and excitement to monetary gain.

Kitty (or Kit, as Michael occasionally called her) was bright, courageous, and decisive. Their marriage was a true partnership, in their social and business life. She and her husband shared all the evidential matters, discussed their respective theories, and solved their cases together, all in the company of Doc. While she did not carry a gun, as both Michael and Doc did, she fearlessly accompanied them into dangerous areas where, clearly, evil lurked.

Only one audio copy of this series has survived from the entire run; it is the concluding episode dated February 6, 1942. This adventure began with Michael, Kitty, and Doc at the waterfront. Kitty talked Michael into checking out a seafarer's auction, while Doc waited with his taxi. During the auction, a dispute over the bidding for a small box resulted in one sailor stabbing another. The attacker fled, jumped in Doc's cab, and they sped away. Meanwhile, Michael and Kitty questioned the victim and a third sailor at the scene. After midnight, Doc arrived at the Piper residence and announced that he took the knife wielder to a deserted barge near an abandoned pier. All three piled in the taxi, drove to the vicinity of the barge, and upon examination of the premises, discovered the corpse of the third sailor, his head bashed in. They notified the police, after they searched for clues around the body.

The next day, the three went to police headquarters and traded theories with Inspector Shield. Then the four of them headed for the hospital and they interrogated the wounded victim again. After the Spur commercial, Michael and Kitty solved the case and directed Shield to arrest the guilty party. As the successful trio took their leave of Shield, the following conversation took place:

MICHAEL: Goodnight, Inspector.

KITTY: Goodnight, now.

SHIELD: Hey, wait a second, where ya going?

MICHAEL: We're taking Doc into protective custody. One murder a night is sufficient.

SOUND: (*Door Closing*)

KITTY: Ah well, what now, Michael?

MICHAEL: Who knows, Kit? Doc, let's get the taxi and go for a drive.

DOC: OK, boss, got some excitement in mind?

MICHAEL: Nooooo, nothing in particular. Let's keep our eyes and ears open. Can never tell when someone will need help and call us in, eh, Kitty?

KITTY: (*chuckling*) I see what you mean, Michael...

DOC AND MICHAEL: (*Chuckling through closing theme music*)

When radio drama evaporated in the late 1950s, Santos Ortego crossed over easily to television, and had principal roles in soap operas, including *The Brighter Day* and *As the World Turns*. There is little documented about the subsequent career of Elizabeth Reller, and none of the standard radio reference books took note of her later years. Ortega was 76 years old when he died in April 1976. His radio associate, John Gibson, outlived him by ten years, dying in 1986 at the age of 81.

Partners In Crime

A number of women sleuths were so involved in the investigative work of their male associate that their crime-solving together approximated partnership. One of them was an actual business partner: Terry Travers in *Results, Inc.* Another was the niece of a retired criminologist, with whom she paired up to solve homicides. Two other pairs got their paychecks from the same employer; one of these couples worked for a newspaper while the other duo were Secret Service operatives. Of the remaining two ladies in this chapter, one was a constant companion to a crime-fighter who had a secret alter ego, while the other one had the distinction of having a boyfriend, who was not her crime-solving boss.

JOAN ADAMS

In the summer of 1944, the NBC-Blue network agreed to air a 15-minute weekly program, in cooperation with the National Safety Council, to focus attention on accident prevention. The series was a crime adventure entitled *It's Murder,* and its two main characters were Rex A. Starr, a retired actor and amateur criminologist, and his niece, Joan Adams, a Broadway gossip columnist. While it may have seemed gracious of the network to mount this show, almost as a public service, they aired it at 11:15 p.m. on Thursday nights, which would have been a difficult time slot to find paying advertisers.

There is some confusion about the length of the time this series was on the air. *It's Murder* debuted on June 8, 1944 and most radio reference books assert it ended on July 6, 1944. However, the sole remaining audio copy is correctly dated August 10, 1944, and the announcer, George Gunn, in discussing the weekly contest, reminded listeners that their entries must be posted before midnight on Monday, August 14th, with the winners to be announced in two weeks, on Thursday, August 24th. It would seem most likely that the series ran through August 1944.

Although this was a program that produced no income, NBC was not reluctant to utilize top flight performers and production personnel. The co-lead, Joan Alexander, portrayed Joan Adams, and this was not the first time her character name resembled her own; she was the voice of Lynn Alexander on the soap opera, *Lone Journey.* As an American youngster, Joan Alexander spent a great deal of time in Europe, where her father owned a linen factory. She studied acting in Europe, as well as in the U.S., eventually getting roles on Broadway and touring stock companies, including parts in *Merrily We Roll Along* and *Mr. Hamlet.*

By the time she was 30, Joan was a successful radio performer; she was in adventure series (*The Man from G-2* and *Perry Mason*), women's daily serials (*The Brighter Day, Rosemary,* and *Young Dr. Malone*) as well as prestigious anthologies (*Columbia Workshop* and *X Minus One.*) In addition, she was Lois Lane in *The Adventures of Superman* and also the secretary of *Philo Vance.* Off-mike, she pursued athletic hobbies: swimming, tennis, and horseback riding.

Her partner at the microphone for *It's Murder* was Edgar Stehli, who was nearly as old as the elderly geezers he played on radio. He was born in 1884, the same year the Civil War general, William T. Sherman, turned down the Republican Party nomination for president with his now-famous statement: "I will not accept if nominated and will not serve if elected." Stehli was 60 when he got the part of Rex A. Starr and he had a long history of network performances. In the early days of *Buck Rogers of the 25th Century,* he was Buck's friend, Dr. Huer, and later he was in several anthologies, including *Best Plays, Lights Out* and *Columbia Workshop.*

The production staff that NBC assembled for *It's Murder* were all successful veterans in network radio. The director, Stuart Buchanan, began in broadcasting doing character parts; Walt Disney fans remember him as the voice of "Goofy" on *The Mickey Mouse Theater of the Air* in the late 1930s. He gradually moved from the microphone to the control booth, directing adventure shows, including *The Falcon,* and soap operas, such as *The Second Mrs. Burton.*

Stedman Coles began his radio writing career toiling in the soap opera factory of Frank and Anne Hummert, as part of the team that put *Mr. Keen, Tracer of Lost Persons* on paper. Later Coles became one of the many authors that produced *The Shadow* scripts. He stayed with the crime drama shows, and in the late 1940s, wrote the scripts for *Famous Jury Trials* and *Roger Kilgore, Public Defender*. Turning out one 15-minute script per week for *It's Murder* must not have been too difficult a task for Coles, despite the fact that the National Safety Council used up at least four minutes per episode with their safety warnings and contest information. So he had the remainder of 11 minutes in which the plot would move from the discovery of the corpse, the interrogation of suspects, and the solution of the mystery. This called for a crisp, taut script and Coles did not disappoint.

The lone surviving audio copy is an episode entitled "Picture Wire Murder." The plot is set in motion when a local member of an artists' colony in Maine was found strangled to death on the beach. Mr. Starr and Miss Adams were among the first to reach the crime scene, which they examined together. She provided Starr with background information on the dead man, they discussed possible theories, and then, without notifying the local law enforcement authorities, the detecting duo set out to question suspects, including the owner of an art supply shop (played by Parker Fennelly).

Joan Adams and her uncle discovered a second body (the wife of the first victim) and determined her death to be a fake suicide. A few minutes later, they unmasked the killer and notified the sheriff. Their rapid solving of the mysteries surrounding the two murders gave announcer George Gunn time to discuss the safety limerick contest for that week and recite the names of the best entries and the grand prize winner from two weeks prior.

Each week the National Safety Council awarded five dollars each to ten runner-up contestants, while a "Grand Prize" of a $ 50 war bond went to the first place winner. Over that summer of 1944, the National Safety Council warnings and limerick contests focused upon different areas of accident prevention: on the farm, in the home, at the factory, in the water, etc. The August 10th program was devoted to the topic of how alcohol consumption imperiled lives of vehicle drivers and pedestrians. In the opening two minutes, Gunn provided statistics on the death tolls of highway fatalities in which alcohol was a factor, and hinted it was hurting America's ability to fight World War II:

ANNOUNCER: Nearly 5,000 Americans might be living
today, and contributing to the war effort, if irre-
sponsible motorists and pedestrians had not yielded
to their thirst while walking or driving.

In the last two minutes of this show, Gunn explained the rules of the safety limerick contest, listed the prizes, and gave instructions on where and when to mail all entries. Radio listeners were invited to furnish a closing line to this limerick:

> Bill decided to take one more nip
> Before driving his car on a trip.
> But that one extra snort
> Cut his trip extra short
> Dah Dahda Dahda Da dip.

Both of the leads in this series stayed active in radio for many years. Stenhi eventually retired from show business late in life. He died in July 1973; he was only one year shy of attaining 90 years of age. Alexander did some television and voiceovers. She was a regular panelist on *The Name's The Same*, an ABC-TV panel show , and was also occasionally on *To Tell The Truth* on NBC, hosted by her old friend, Clayton "Bud" Collyer. Joan was selected to reprise her role as Lois Lane when *Adventures of Superman* was brought to television by CBS in a weekly cartoon series in the late 1960s.

TERRY TRAVERS

There were only two series in the history of dramatic radio with a lady crime-solver in which the co-leads were both prominent Hollywood movie stars. The first one was *Miss Pinkerton, Inc.* in 1941 with Joan Blondell and Dick Powell and the second was *Results, Inc.* with Claire Trevor and Lloyd Nolan, which began three years later.

Lawrence E. Taylor created *Results, Inc.* and the first program aired on October 7, 1944. It was a sustained series on Mutual with a very short run; the last episode was broadcast on December 30, 1944. Of the dozen half-hour weekly shows that aired, only two audio copies have survived, the first and last episodes.

Despite its short run, the production staff kept changing. The earlier scripts were written by Leonard St. Clair and Bill Hampton, but their places were later taken by Sol Stein and Martin Wirt. Felix Mills, whose orchestra was heard on dozens of radio shows, including *Burns and Allen, The Man Called X, Chandu the Magician,* and the children's' classic, *The Cinnamon Bear,* was originally responsible for the music on *Results, Inc.,* but then Russ Trump's orchestra took over. Henry Charles began as the

Edward G. Robinson, Claire Trevor and Humphrey Bogart in
The Amazing Dr. Clitterhouse (Warner Brothers, 1938)

announcer for the show, but later he was replaced by Bob O'Connor. However, the series had the same producer for the whole three months, Don Sharpe. He was very experienced, and, at different times, also produced: *Dangerous Assignment, Adventures of Michael Shayne, Four Star Playhouse,* and *Richard Diamond, Private Detective.*

Both Claire Trevor and Lloyd Nolan were established film stars when they accepted roles in radio's *Results, Inc.* Trevor, who was born Claire Wemlinger in 1910 in the Bensonhurst section of Brooklyn, began acting in stock companies in the late 1920s to help support her family, after the failure of her father's clothing business. This led to parts on Broadway, and later, screen roles in Vitaphone movies that were filmed in Brooklyn. By 1933, Trevor was in Hollywood, working her way up the ladder as a contract player, usually playing saloon gals, gangland molls, and assorted tough ladies. In 1937 she received her first Academy Award nomination as Best Supporting Actress for her superb performance in *Dead End.* Another of her earlier successes was 1939's *Stagecoach,* with John Wayne, in

John Wayne, Claire Trevor in *Stagecoach* (United Artists, 1939)

which she played a woman of easy virtue, a role that should have garnered her another Oscar nomination, but it did not.

When she joined Lloyd Nolan at the radio microphone, she was a
highly respected film star under contract to RKO Studios. Nolan had followed a somewhat similar and successful trail from the stage to Hollywood.
He was born in 1902 in San Francisco, studied at Stanford, and was getting

good stage roles by the time he was 25. Nolan migrated south to Los Angeles and found plenty of movie work, often cast as a tough detective (including *Michael Shayne)*, a soldier, or news reporter. In the mid-1940s, like Trevor, he was making about five full-length movies each year.

Neither Trevor nor Nolan had much broadcast experience and it showed in their radio series, which was sprinkled with minor flubs by the co-leads. This was made more obvious by the flawless performances of the veterans in the supporting cast: Vic Ryan, Bea Benaderet, Joe Kearns, and Wally Mayer. *Results, Inc.*, described by its creator as a comedy-mystery, took its name from the small business run by Johnny Strange and his only employee, a secretary, who would later be designated Vice President. This was not a detective agency, but an all-purpose service company that would solve any problem, find any missing person or object, and perform any task. Of course, the visit of any client to their office always led to some criminal activity, which Strange and Travers would resolve.

The debut episode, "Haunted House," began when Johnny placed two classified ads, one announcing his new business venture and a second one advertising a vacancy for a secretary. Theresa Travers, who Johnny promptly dubbed "Terry," answered his want ad, and the two briefly described their respective employment history. She had been, in succession, a newspaper reporter, a lingerie model, a magician's assistant, and a secretary to a showbiz producer. Johnny told her he had started as a trombone player in a circus band, then became a news correspondent, served as a deck hand on a freighter to South America, and was briefly a detective before starting this new business.

By the time they agreed on her salary (25 % of all commissions, plus bail money and health benefits) their first client walked in: a mystery novelist who wanted them to find her a haunted house for atmosphere. The next day, they located a suitable abandoned building, outside the little village of Ashton Junction. Since Terry was a camera hobbyist, she took photographs in and around the scary mansion, which later revealed it was the current hideout of Eddie Carson, a homicide fugitive who had escaped from an asylum for the criminal insane.

Back at their office, Johnny wanted to telephone the police with the news of their discovery, but Terry overruled him.

TERRY: What are you going to do?

JOHNNY: Phone the police—

TERRY: Wait, we've got something big here. The police all
            over the state have been searching for Carson. Now,
            if we were to capture him—

JOHNNY: Now, honey, you shouldn't drink that film developer.

TERRY: Oh listen, you big galoot; it would make us! Think
            of the publicity! Eddie Carson captured by the
            President of Results, Inc. How would that look on
            the front page?

JOHNNY: Hmmm, you know, sweetheart, I think you've got
            something.

That night they drove back to the haunted house, but Johnny's car ran
out of gas. Since this took place in World War II, there was some discussion
about his book of gas rationing stamps. The intrepid duo were forced to walk
the rest of the way to the house, and enroute they were met by a volunteer air
raid warden who told them a practice blackout would start soon.

Inside the killer's spooky hideout, Johnny's carelessness led to their
capture by the maniac. Terry, a resilient and resourceful woman, tricked
the murderer into posing for some flash camera shots, which alerted the
air raid warden. His arrival distracted the crazed killer and Johnny knocked
him out. Case closed.

The last episode of the series, "Mummies Walk," aired on December
30, 1944, so the scriptwriters could not resist having the final adventure
take place on New Year's Eve. Just three months after her hiring, it was
apparent that Terry, though technically still a secretary, was now a full
partner in their two-person agency since Johnny referred to her twice,
neither in jest, as their Vice President.

The plot was set in motion when a local museum watchman arrived
at their office and told them his tale of mummies that walked. Johnny was
convinced the case was not grounded in reality, but Terry prevailed and
they went to the museum. Later that evening, they confronted two thieves,
who knocked Johnny unconscious and kidnapped Terry in a mummy's
sarcophagus. The two adventurers were later reunited at the hideaway of
the larcenists, where, after Terry created a suitable diversion, Johnny subdued the evildoers.

Mutual broadcast this series for three months but failed to find a
sponsor; by the end of the run, they were airing public service announce-

ments, such as appeals for enlistments of nurses in the Army Nurse Corps. It's also possible that the movie obligations of Trevor and Nolan pulled them away from the low wages that radio paid in those days. In either case, this lack of success on network broadcasting did nothing to slow the progress of either star in their entertainment careers. Both would go on to appear in over 60 motion pictures in their lifetimes.

Trevor continued her busy film schedule for the next 15 years. She won the Oscar for Best Supporting Actress for her portrayal of Edward G. Robinson's alcoholic ex-mistress alcoholic singer in 1948's *Key Largo,* with Humphrey Bogart. Six years later, she was nominated for another Academy Award as a supporting actress in *The High and the Mighty,* with John Wayne in the lead. Trevor's television career was not as extensive as Nolan's, but she won an Emmy in 1956 for her role in "Dodsworth," a 90-minute show aired live on NBC-TV's *Producer's Showcase.*

She retired to Newport Beach, California to enjoy many years of painting, charity work, and travel, accepting only infrequent guest shots on television, including *Love Boat* and *Murder, She Wrote.* She was a dedicated philanthropist, and in 1999, donated $ 500,000 to the University of California's School of the Arts (in Irvine), which was renamed in her honor. Trevor died peacefully in April 2000, just one month after her 90th birthday.

Nolan worked continually in the movies, on stage, and on television for 25 years after World War II. He was a regular on several television shows, including *Dupont Show of the Week, The Untouchables, Outer Limits,* and was one of four actors to play *Martin Kane, Private Investigator.* In one of his more popular roles, Nolan was Dr. Chegley, on *Julia* where he costarred with Diahann Carroll, who played his nurse. This series, which began in 1968, was the first program on American television to have an African-American woman in a leading role, who was not a maid.

On the stage, Nolan won critical acclaim as Captain Queeg in *The Caine Mutiny Court-Martial ,* and later he reprised the role on television in 1955 on *Four Star Jubilee,* for which he won an Emmy. He retired in California in the late 1970s and died in 1985. His small marble headstone over his grave at Westwood Memorial Park in Los Angeles reads simply:

LLOYD BENEDICT NOLAN
1902-1985

MARGOT LANE

Radio's three most famous program openings, indelibly etched into American culture, are the distinctive introduction to *The Lone Ranger,* the door creaking open to signal *Inner Sanctum Mysteries,* and an organ playing the theme of Saint-Saens' *Omphale's Spinning Wheel* while a compelling voice uttered: "Who knows what evil lurks in the hearts of men? The Shadow *knows!*" followed by chilling laughter.

The Shadow, created in 1930 almost by accident at Street & Smith Publishing, would progress in stature rapidly, eventually to become one of their greatest fictional characters. His phenomenal success in radio, detective magazines, pulp novels, movies, and even comic books, was beyond anything the publishers imagined from his origin: a mysterious radio host who introduced the adventures of different crime-fighters, taken from the pages of *Detective Story* magazine.

In the early thirties *The Shadow* gradually evolved, both on the printed page and in his radio programs, each influencing the other, with the radio character becoming the version that more of his public accepted. The vicious, black-cloaked avenger from the pulps, with a blazing semi-automatic in each fist, gave way to the staid, suave, Lamont Cranston, and his alter-ego, the invisible Shadow, neither of whom carried firearms. Although the pulp's Shadow , written by Walter B. Gibson, had at least a dozen confederates, including Burbank the radio operator, two agents named Harry Vincent and Cliff Marsland, and a criminologist called Slade Farrow, radio's Shadow made do with very few.

CBS began airing this character on network radio in July 1930, however the series was not named *The Shadow* until January 1932; Margot Lane would arrive five years later. *The Shadow* remained on network radio for over two decades, and despite the horror-filled melodrama of most of the story lines, the series received strong ratings throughout nearly a quarter century.

There were only three supporting characters on *The Shadow*; one later disappeared, one changed completely, and the third became the co-lead. Shrevie, a loquacious taxi driver, was originally the primary transportation for the Shadow, but he was dropped from the series after a few years. Commissioner Weston, in the early years of the program, was a blustering, arrogant, police official who tried to embarrass or confound the Shadow. By the mid-40s, Weston and Cranston were the best of friends and Weston both respected and admired the Shadow. The third character,

Agnes Moorehead, "Margot Lane" (*The Stumpf-Ohmart Collection*)

not added until 1937, was usually described as "Cranston's constant companion, the lovely Margot Lane," and she moved up from cordial friend
status to co-lead as Cranston's lover, but only in the most chaste of rela-
tionships.

This lady, "the only person who knows to whom the voice of the
invisible Shadow belongs," obtained her first name from Margot Stevenson,
the Broadway girlfriend of an early producer of *The Shadow,* who wanted
her for the part, but it went first to Agnes Moorehead. (However, Stevenson

would become the second woman to play Margot Lane.) In knowing the Shadow's secret identity, Margot was in a rare position of trust. On network radio, Lois Lane did not know Clark Kent was *Superman*, Mesquite Molly did not suspect Steve Adams was *Straight Arrow*, and Joan Mason was unaware that Patrolman Dan Garrett was *The Blue Beetle*. Perhaps the only lady to equal Margot's special privilege would be Barbara Sutton, who knew that underneath the cowl and cape of The Black Hood was her boyfriend, Kip Burland (see chapter 4.)

Because of its longevity, 1930 to 1954, a multitude of radio performers got an opportunity to play the co-leads on *The Shadow*. We know that at least ten men played the Shadow at different times: James LaCurto, George Earle, Robert Hardy Andrews, Jr. (all prior to 1931), Frank Readick, Jr. (1931 to 1935), Carl Kroenke (mid-'30s), Orson Welles (1937 to 1938), Bill Johnstone (1938 to 1943), Bret Morrison (1943 to 1944 and 1945 to 1954), John Archer (1944 to 1945) and Steve Courtleigh (five weeks in 1945.) Playing Margot at the microphone were eight different women, beginning with Agnes Moorehead and Margot Stevenson (1937 to 1939), a followed by Marjorie Anderson (1939 to 1944), Judith Allen (1944 to 1945), Laura Mae Carpenter (three months in 1945), Lesley Woods (1945 to 1946), Grace Matthews (1946 to 1949) and Gertrude Warner from fall of 1949 until the series ended in December 1954.

The combined total of writers, at least twenty, was a very large number, even for a long running show which The *Shadow* definitely was. These writers sometimes worked alone, and sometimes as part of a team, and as a result of many hands at the typewriter, Margot's involvement in Cranston's investigations varied from year to year. Sometimes she was an interested bystander to his crime solving inquiries, other times she was an equal partner, and infrequently she was a helpless victim, awaiting rescue by Cranston or the Shadow. But despite the changing nature of her participation in this crime drama from one period to another, she remained intelligent, valiant, steadfast, and self-confident.

The Shadow was probably the only fictional radio crime-fighter to have his network scripts published in a high school textbook by a prominent firm…twice. In 1954, and again in 1977, Scott, Foresman & Company, the premier publishing company for scholastic books, issued two volumes, both titled *The Shadow Knows*, which contained sixteen complete radio scripts. The breadth of these scripts gives us a detailed insight into Margot Lane's character, behavior, and accomplishments.

She was both witty and literary-minded in "The Blackball Murder," as she teased Lamont, at the oars of their small boat, by paraphrasing Shakespeare, "Row on, Macduff." In the episode, "The Horror in the Night," the two were in a deserted inn, festooned with dust and cobwebs:

LAMONT: This place could use a woman's touch.

MARGOT: It could use a dozen women's touches—all scrubwomen.

As Lamont's constant companion, in the adventure called "Heartbeat of Death," she demonstrated not only initiative and invention but also the talents of a radio sounds effects artist. She was handed some terse, written instructions by Cranston, which she successfully fulfilled by locating a gourd and secretly tapping out heartbeats near the air registers, thus tricking a killer into a confession.

There were even occasions when Margot's detective skills were a match for those of Cranston. An excellent example of this talent was revealed in "The Case of the Dead Man's Shoes" when Cranston sent her to interrogate a suspect, Leon Selby, and she cracked the case:

SELBY: What are you looking at, Miss Lane?

MARGOT: That glove, Selby, that long black suede glove.

SELBY: Glove? Oh, that…that belonged to Mr. Windsor's last wife, I believe.

MARGOT: What's it doing out on the desk?

SELBY: Probably been there for weeks…

MARGOT: Really? That's funny. It's lying on top of a theater program, and a couple of ticket stubs dated yesterday. Must have been placed there after…

SELBY: Don't get any strange ideas, Miss Lane.

But it was too late for Selby. Margot correctly deduced that Ellen Blake and Mr. Windsor were at the theater matinee when the murderer returned the missing necklace to the jewelry store after he murdered Archie, so the killer must be Selby.

Throughout their hundreds of cases, Margot found clues that Lamont missed, gained guilty admissions from suspects based upon her incisive interrogation, and utilized her skill in acting and disguise to complete a formidable undercover operation. In "House of Horror" which aired November 17, 1940, Margot impersonated a victim of a jewelry loss so that a fake fortune-teller led her to a kidnapper. Both Cranston and Margot went undercover during "The Master of Torture" episode, which took place in Athens. To gain the trust of an American gangster who had been deported to Greece, Cranston pretended to be an ex-con named "Lefty Lamont" with Margot as his gun moll. In that foreign adventure, while Lamont was busy elsewhere, Margot was able to sneak into the gangsters' hideout by herself, although she was unable to escape before the crooks returned.

The Shadow had an advantage over most of the more realistic crime dramas, since once you bought the premise that Cranston actually had the power to cloud anyone's mind so they could not see him, nothing else in the story line would seem too outlandish. That freed Lamont Cranston and Margot Lane to encounter a mad scientist who transposed the brainwaves of a gorilla into a woman, a crazed subway conductor who traveled with the corpses of his victims in a ghost train beneath a carnival, and an evil doctor whose formula, when injected into a human, put them in suspended animation as wax dummies. But through it all, Lamont and Margot remained calm, decisive, and rational with no diminishing of their deductive powers.

"The Altar of Death," which was broadcast March 15, 1942, took place in the tropics where the two investigated a series of mysterious deaths and they determined that a white man had convinced the natives on a volcanic island that he was a god and they must honor him with human sacrifice. Margot began the unraveling of the mystery when she noticed the belt buckle of one victim being worn by their chief suspect:

MARGOT: Lamont, did you notice the initials on the silver
            buckle he was wearing?

LAMONT: No, I didn't notice.

MARGOT: They were capital "M," small "ac," capital "B."

LAMONT: Hmmm, capital "M," a...c, capital "B."

MARGOT: And didn't Owney tell us that the first two people
            to vanish from San Luis were a Mr. MacBeard and
            his wife?

LAMONT: Yes, Margot, I'm afraid he did...

Another advantage *The Shadow* radio series had was that neither Lamont Cranston or Margot Lane had any family, employment, or financial obligations, so there were no limitations on their time or travel. Cranston was customarily described as "a wealthy young man about town" and Margot Lane was "his constant companion," but this was the extent of their respective livelihoods. Anthony Tollin, an authority on this radio series and the author of the book, *The Shadow: The Making of a Legend,"* commented on the couple's lifestyle:

> "Sometimes she was a reporter, and occasionally Cranston was an author or playwright, but mostly they were just wealthy dilettantes with too much spare time on their hands."

Tollin also makes it clear that Cranston's "constant companion" had spunk, valor, and independence, even on her early years in the series.

> "Margot was far more a full associate and less of a damsel-in-distress during the 1937-1939 seasons under the script supervision of Edith Meiser, the writer-director who also brought Sherlock Holmes to radio. Margot rescued Lamont then as often as she needed rescuing (driving her car through the window of a burning building in 'The Firebug.') Meiser was a former suffragette who prided herself in being arrested at two separate women's rights demonstrations some sixty years apart so her female characters were not dainty things requiring constant rescuing. The lovely Margot Lane didn't learn to be a victim until Edith's departure from the series."

Another expert on The *Shadow,* who almost shares his name, Karl Schadow, points out that even the Margot of the 40s and 50s made courageous and vital contributions to the solution of the mysterious crimes in these later adventures:

> "While the scripts of the later seasons are not as intriguing or exciting as those of the earlier years, Margot still plays a pivotal role in many of the 1940s and 1950s formula-ized

episodes. Her helpful suggestions concerning smelling salts to revive herself after viewing a mutilated corpse, alerts Lamont that the lack of leopard scent may be a key to solving the grisly murders in 'The Leopard Strikes.' Later in this story, her keen eyesight locates the residence of the cult responsible for the killings. Her notion regarding the activities of a dangerous gang being able to actually see every move of the police, reminds Lamont of a professor's inventions in 'Death to The Shadow.' Margot puts herself in grave peril after being admitted as a patient in a sanitarium in 'Death In A Minor Key,' while seeking the advice of Ahmed in 'The Crystal Globe,' and as the newly employed maid of the vicious Bailey household in 'The Chill of Death.' Her nightmare of murdering a countess leads to the apprehension of a quack in 'The Dreams of Death.' In 'Death and the Easter Bonnet,' the adornment that she just purchased involves international espionage and brings her a $500 reward following the recovery of a stolen government secret. Margot poses as a newspaper reporter in 'The Vengeance of Angela Nolan' to acquire information from a suspected murderess."

The Shadow, on radio, literally went around the world. The series was broadcast in Australia, through Grace Gibson Productions, with an Aussie cast headed up by Lloyd Gamble. Other English versions aired in South Africa and Great Britain, while 250 episodes in Portuguese were heard in Brazil and about the same number in Spanish, aired in Mexico. Today, approximately 245 audio copies of *The Shadow* episodes in the U.S. are currently being traded by collectors and sold by nostalgia dealers.

On the silver screen, *The Shadow* has had several American productions, beginning in 1931 with six short "filmettes" based upon the short stories in the pulp magazines. In 1937 the first full-length film of *The Shadow* was released; it was a Grand National movie, *The Shadow Strikes,* which starred matinee idol, Rod La Rocque, in the lead as Lamont "Granston." His constant companion made her first screen appearance when La Rocque was brought back in Grand National's 1938 *International Crime;* this time they bungled her name, calling her Phoebe Lane, instead of Margot. It may have been for the best, since Cranston and Lane were merely supporting characters in this motion picture.

Columbia Pictures released in 1940 a new 15-chapter serial, *The Shadow,* with craggy-faced Victor Jory as Cranston/Shadow and Veda Ann Borg playing Margot. Curiously, Jory spent a lot of time in this serial disguised as Lin Chang, trying to infiltrate an organization called "The Black Tiger." Monogram briefly had the movie rights to *The Shadow,* and they quickly churned out three B-pictures in 1946 based on that character: *The Shadow Returns, Behind the Mask,* and *The Missing Lady.* Handsome and athletic Kane Richmond, who had also portrayed *Spy Smasher* and *Brick Bradford* in the serials, got the lead of Cranston/Shadow in this trio. Barbara Reed was his Margot.

No motion picture concentrated on the invisibility that was so useful to radio's *The Shadow* until the 1958 Republic release, *The Invisible Avenger.* The star was Richard Derr, and as *The Shadow,* he could not only cloud men's minds, he also received hypnotic messages from a mystic named Jogenda. The most recent appearance of the master of darkness in movie theaters was in 1994 with Alec Baldwin as Cranston/Shadow and Penelope Ann Miller as the lovely Margot Lane. This colorful, but disappointing, picture was based substantially on the character of the pulp magazines and its scriptwriters apparently had never heard of *The Shadow* on radio.

Mutual's termination of their network's *The Shadow* on the day after Christmas 1954, did nothing to diminish the presence of this revered character in our popular culture. For the next fifty years, rebroadcasts of *The Shadow* continued on stations throughout the country. Nine new novels of *The Shadow* were published, beginning in 1963, by Belmont Books, and the next year, *Archie Comics* brought him back to the comic book newsstands for a short period. The new novels, as well as paperback reprints of those from the '30s and '40s, continue to sell briskly to the present day.

Patsy Bowen

Throughout the world, Nick Carter and Sherlock Holmes are two of the best known fictional detectives whose exploits have been in various media for over a century, including books, magazines, motion pictures, television and radio. However, Carter's assistant, Patsy Bowen, was at his side only on radio and in the comic book pages.

With over 500 novels, and about a thousand short stories, beginning in 1886, Nick Carter may have the longest tenure of any sleuth in this nation's crime fiction. Throughout the years, he has changed significantly in all aspects of physical size, age, associates, and employment so that the only common denominators of all versions of Nick Carter were his name and his desire to thwart evil.

Nick Carter, originally a young loner nicknamed "The Little Giant," was created by author John Russell Coryell and literary representative Ormond G. Smith; the latter was the son of one of the original founders of Street & Smith Publishing. For his initial seventeen years, this pint-sized crime-buster had most of his adventures chronicled in the dime novels, most of them written by Frederick M. Van Rensaler Day. After that author's death, a host of Street & Smith writers, under house pen-names, continued churning out the novels and stories. By the turn of the century, Nick Carter was as popular in the Street & Smith pulp magazines as he was in the dime novels.

Carter used his diminutive stature, ingenious disguises, and great physical strength to solve crimes and bring evildoers to justice. He was so adept at switching from one disguise to another, he could have been on the stage as a quick change artist. In one adventure, while on surveillance, he was disguised as an elderly Polish matron when he stepped into a doorway for a few moments and emerged as an African tradesman. In addition to his great courage and scientific knowledge, Carter was an accomplished linguist who could fluently speak nearly every foreign language, as well as lip-read the conversation of anyone, regardless of their ethnic origin.

In the dime novels and pulp magazines, he gradually acquired the services of two associates: Chickering Carter, whose nickname was Chick, and Paddy Garvan, who was sometimes called Patsy. Chick was Nick's editor and confidant, but although they shared the same surname, they were not related. Garvan was a teenaged bootblack and his street contacts were of help to Nick. Many years later, on network radio, these two characters would evolve into Chick Carter, Nick's nephew, and Patsy Bowen, his lovely assistant.

By the early 1900s, Nick Carter became a popular subject in the emerging film industry, predating the success of Street & Smith's other famous character, *The Shadow* who debuted in 1930. From 1908 to 1917 a number of silent motion pictures about Nick Carter were filmed in the U.S. and Europe, particularly France. In 1939 the first film by a major U.S. studio featuring the famous detective, titled *Nick Carter, Master Detective,* was released by MGM. In this movie, The Little Giant had become a tall, urbane, middle-aged sleuth who used his brain more than his brawn, and who would never disguise himself as an elderly Polish matron.

Canadian-born Walter Pidgeon played Nick Carter, but none of his associates mentioned in the novels or pulps had made it into the film. Instead, a new character, Beeswax Bartholomew, a beekeeper, who resembled a mild and eccentric Dr. Watson, was Nick's primary sleuthing buddy; he was portrayed by Donald Meek. Two other actors in this movie are of interest. Wally Maher, who played Cliff Parsons in this 1939 film, would go on to substantial success in radio drama, with major roles on *The Adventures of Michael Shayne, I Love a Mystery,* and *The Whistler.* The part of Dave Krebs was played by Milburn Stone, who, sixteen years later, would become Doc Adams on television's *Gunsmoke.*

The film version of Nick Carter was more of a counterspy than a private detective; he was assigned to foil enemy spies who were trying to steal secret plans for a new fighter at the Radex Airplane factory. Although only moderately successful at the box office, MGM quickly released two more the next year, again with Nick and Beeswax. These were *Phantom Raiders* (later retitled *Nick Carter in Panama)* which took place on an ocean liner, and *Sky Murder,* in which the setting was a transcontinental clipper. Part of the reason for the mediocre audience response to these three films may have been that Donald Meek, pulling live bees out of his pockets, was more interesting to the viewers than was the master detective, searching for clues.

It was also in 1940 that Street & Smith came out with their own comic book line, using their popular characters from the pulps, several of whom would later become well known radio heroes: *The Shadow, Frank Merriwell, Doc Savage,* and, of course, *Nick Carter.* The latter made his first appearance in the March 1940 issue of *Shadow Comics* but he more resembled his movie counterpart than his pulp magazine persona as he was tall, urbane, gray at the temples, and had a pipe clenched in his teeth. Nick would continue to appear in comic book form regularly until 1949, and about the time his radio show was launched in April 1943, Patsy Bowen began sharing his comic book pages. The Street & Smith artists drew her as an attractive blonde.

In addition to Patsy, radio's Nick Carter had, on his half-hour show for Mutual, some other associates created for this network series. Scubby Wilson, a newspaper reporter, and Sgt. Riley of the local police; the latter was replaced by Sgt. Mathison (or Matty) by the end of the first year. A third associate, Penny, a street-smart guy and dedicated helper of Nick's, lasted about a year and then was dropped. The network called its series

The Return of Nick Carter, although he had never left, but that inappropriate title was used only until December 1946 when it was changed to *Nick Carter, Master Detective.*

Lon Clark, who was born 1911 in the village of Frost, MN, won the part of Nick Carter, and kept that role for the full twelve years it was on the airwaves. He had begun his career on the stage, as both a singer and an actor. Clark was part of the *Cincinnati Opera* when he began finding radio jobs in Ohio, which quickly led to major network roles in New York on *Wilderness Road, Bright Horizons,* and *The Mysterious Traveler.* Clark, a handsome and engaging fellow, became so popular as the master detective, that his picture was featured on the cover of *Nick Carter* comics.

The initial Patsy Bowen was Helen Choate, whose radio background was primarily in the New York soap operas (*Pretty Kitty Kelly, Rosemary,* etc.) although she also had a continuing role as Jane in *Eno Crime Club.* Choate played the role of Nick's assistant until 1946 when Charlotte Manson took her place, holding the role until August 1955 when it went off the air. Manson, a Manhattan native, was six years younger than Lon Clark; she also began her career on the stage. She had leading roles in Shakespeare productions, Rosalind in *As You Like It* and Ophelia in *Hamlet* before she found her niche in radio. Manson started with NBC in 1939 on *Parade of Programs,* and then found roles on several soap operas; she eventually became the Ronson Lighter Gal on that superb game show, *Twenty Questions.*

Unlike most crime-sleuthing couples on radio, Patsy and Nick, while certainly in a close, personal relationship, were not in love. Patsy's steady boyfriend was Scubby (the voice of John Kane), who often began a sentence with "If I didn't love you so much...." They never made it to the altar, though Scubby did propose marriage from time to time. As for Nick's romantic aspirations, he followed the practice of most of his fellow radio sleuths and remained a perennial bachelor.

Since this series was on the air for over a decade, director Jock MacGregor used many writers over the years, including Alfred Bester, Barth Conrey, Norman Daniels, Max Erlick, Ferrin Fraser, Dave Kogan, Milton Kramer, John McGeevey, and Jim Parsons. These were in addition to MacGregor, who as the director, wrote every third or fourth script. Despite the variety of authors who worked on this radio series, the plots remained pedantic, hackneyed, and frequently unrealistic. Some of the scripts sounded as though the writers were mired in the dime novels where Nick had originated.

Occasionally an incident would appear in a radio episode that seemed lifted intact from the 1800s printed adventures. In "The Drug Ring Murder: or The Mystery of the Left Handed Killer," which aired November 10, 1943, Nick was captured by two thugs. They tied his hands behind him, strangled him with ropes, and leaving the ropes binding him, shoved him into a canvas bag, and took him into the bay in a speed boat. The two criminals threw their victim into deep water and then roared away to their yacht. Minutes later, Nick, dripping wet, confronted the devious duo. The master detective later explained that he had the power to expand his neck and wrists, therefore when he relaxed them in the bag, the ropes fell off. When he was thrown into the water, he cut his way out of the bag, swam to the surface, grabbed the side of the small boat, and the criminals unknowingly took him to their floating hideout.

In another episode, radio's Nick Carter duplicated the disguise skills of his pulp magazine predecessor, as well as his linguistic prowess. Nick had disguised himself as a ragtag beggar, and with a changed voice, passed himself off as Snuffy Davis. When he tried to mooch some coins from Patsy on the street, even she did not recognize him. Then in a later adventure, trapped in the basement hideout of "The Secret Six," Nick, without even meeting their South American butler, impersonated the voice of the butler so well, he fooled the entire gang. Nick, very skilled with locks, as was his pulp magazine ancestor, picked at least one lock in every other episode, usually illegally.

The scriptwriting did not seem to improve much over the years. The January 26, 1947 program was introduced by the announcer in phraseology that sounded akin to what a kid might read in a comic book:

ANNOUNCER: Our story begins in a hideout in a deserted riverfront warehouse where a masked man, whose confederates fear and know him only as The Boss, watch him silently as his lieutenant, Big Louie Arkin, opens a large bag of currency and dumps it on the table.

The program opening changed a few times over the years. During the first three years listeners heard:

SOUND: (Five rapid knocks on door...PAUSE...five more rapid knocks)

SOUND: (Door opens quickly)

PATSY: What's the matter? What is it?

MAN: Another case for Nick Carter, Master Detective!

In August of 1946 Cudahy Company took over the sponsorship of the show and they decided that Nick should share the opening with their household cleaner, so for the next three years, the beginning sounded like this:

ANNOUNCER: Old Dutch Cleanser, famous for chasing dirt
Presents Nick Carter, famous for chasing crime.

It was also about this time that they disbanded the custom of declaring a double title for each episode, such as "The Substitute Bride: Or The Mystery of the Night Ferry."

While the writers frequently made Nick resemble his dime novel ancestor, Patsy was clearly a modern woman: independent, self-confident, and courageous. As many of her female compatriots on other detective shows (Margot Lane, Jean Abbott, etc.) were made into victims to await their rescue by the hero, Miss Bowen infrequently found herself in dangerous spots from which only Nick could extricate her. But she was more likely to be at his side, finding clues, confronting suspects, and even getting confessions from guilty parties. In one adventure, she and Nick were with two others in a mansion and Patsy was the only one to notice a gunman at the window. Her warning, before the shots rang out, saved the lives of four people.

Patsy improved her deductive skills over the years. In a November 1947 episode, "The Barefoot Banker," she and Nick determined that whoever had driven a particular car the prior night was the killer. Only two people had access to the vehicle, a tall man who smoked cigarettes, and his shorter, non-smoking sister. While Nick was busy with an interrogation, Patsy examined the automobile in the garage. When she found no cigarette butts in the ashtray, she was willing to eliminate the brother as a suspect. She then checked the rear-view mirror above the windshield, which was correctly adjusted for Patsy's view. Since Patsy was the same height as the sister, Patsy correctly concluded the sister had driven the car last, and therefore was the killer.

Occasionally the scriptwriters veered off into areas that resembled bad science fiction, but even those programs showed Miss Bowen's courage and independence to good advantage. She was not afraid to overrule Nick's infrequent poor decisions. A good example is heard in the February 9, 1944 adventure, "Mystery of the Vampire Killers" where five women are attacked in a local park, three of whom are killed by child-sized creatures who drop from the trees and drain the victims of their blood. Carter sets a trap for the vampires, using himself as bait, but he is instead knocked unconscious by the tiny terrors who escape. The master detective concludes that the tree vampires only victimize women so he decides to dress up in lady's clothes and return to the park that night. Patsy confronts him:

PATSY: There is one other way, Nick, a much better way.

NICK: Hmmm, what's that Patsy?

PATSY: Use me for bait.

NICK: Nothing doing, Patsy. You think I'm going to risk your life?

PATSY: You risked your own life. I'm no better than you are.

NICK: Well, that's different. No, I won't let you and that's final.

PATSY: It's not final, Nick. Three girls have been murdered already and two others attacked. If I can help put an end to these killings, by serving as bait for these awful things, I want to do it.

NICK: But, Patsy...you—

PATSY: No, Nick, no but's about it. I want to do it. You've got to say yes. I'm serious.

Nick reluctantly agreed; she acted as the bait and the creatures were captured. They turned out to be "Asian pygmy women from the Tukk Tribe in China" and their sect required a blood sacrifice within six months if their husbands died. All of them were on tour in the U.S. during the six months after their husbands died, so they went into the park to get fresh blood for their sacrifice.

By the fall of 1946, Miss Bowen had become so proficient in the detective craft, she solved a murder mystery case entirely by herself. "The Red

Goose Murder" was aired on September 1, 1946 and Lon Clark did not appear in the entire show since "Nick's out of town." This adventure began with a call to Nick's office regarding a mysterious death of a singer at the Red Goose Club. Patsy offered to help, canceled her dinner date with Scubby, and went to the club. Scubby and Sgt. Mathison met her at the club crime scene and their investigation ensued. She found all the significant clues and evidence, including the guitar string used to strangle the victim, and she determined the time of death and motive. Despite her protests, Mathison arrested the wrong man and took him to the station. Patsy and Scubby returned to the manager's office where she revealed to him all the evidence she had gathered, which pointed to the manager as the killer. When he threatened her and Scubby, she pulled a pistol from her purse.

MANAGER: Awright, so what if I killed her? I'll have you two taken care of fast—

PATSY: Sit down, Mr. Bradley.

MANAGER: You can't scare me with that little popgun.

SCUBBY: Don't kid yourself, Mr. Bradley. Patsy knows how to use that gun, and she will if she has to. And a .22 makes just as good a hole in a man's heart as a .45 does, if it's aimed right—the way Patsy aims.

Radio Best magazine followed the practice of *Radio Mirror* in a monthly pictorial summary of one episode from a popular radio show. The November 1947 issue featured *Nick Carter, Master Detective* in "the Case of the Blue Mink." Nick and Patsy were investigating fur thefts from a local department store. Patsy went undercover as a clerk and saw a customer steal a blue mink coat. Although the thief escaped, she dropped her theatre tickets, which Patsy retrieved. Patsy went to that theatre and saw another woman wearing the stolen mink. At intermission, Patsy secretly reclaimed the stolen fur, but was discovered by the crooks and taken at gunpoint to their hideout. In the last photograph, Nick had just rescued Patsy from the fur thieves.

Nick Carter, Master Detective lasted until September 1955, long after most radio network drama had been canceled. A little over 100 episodes are currently in circulation, most from the period between 1946 and 1950. Nick Carter's appeal has continued. In 1972 there was a made-for-television movie, *The Adventures of Nick Carter,* which starred Robert Conrad, fresh from his five-year successful television run on *The Wild, Wild West.* There were additional motion pictures featuring the master detective that were filmed in Czechoslovakia and France. An animated cartoon series of *Nick Carter* aired on Italian Television from 1971 to 1976. Paperback novels, touting a new Nick Carter, whose accomplishments are in the realm of James Bond, have been published here in the last ten years.

After their radio series ended, Clark and Manson stayed active in their craft. She did television shows (including *The Untouchables*) and commercial voiceovers. She had been retired for about ten years when she died in December 1996 at the age of eighty.

Clark did some television work but concentrated on his first love, the theatre. He stayed in the New York area, and in 1956, was chosen to take the Broadway role that Jason Robards vacated: the father in Eugene O'Neill's *Long Day's Journey Into Night.* Clark also taught poetry and Shakespeare at Adelphi University on Long Island. He enjoyed meeting his longtime fans while attending vintage radio conventions, including the Friends of Old Time Radio in Newark, NJ. Clark was 86 when he passed away in Manhattan in October 1998.

ANN WILLIAMS

A jazz-loving photographer and crime-solver, who began in the mystery novels of George Harmon Coxe, evolved into a radio hero through the scriptwriting of Alonzo Dean Cole. *Flashgun Casey* began on CBS in July 1943, and under a sequence of slightly different titles, would air on network radio for nine of the next twelve years. Except for the first few months, Staats Cotsworth played Casey for the entire run, however five different women played the co-lead of Ann Williams.

This series' title was changed to *Casey, Press Photographer* in 1944. The following year it was retitled *Crime Photographer* and this name lasted for two more years. CBS expanded the series title to *Casey, Crime Photographer* in March 1947 and kept that intact until the series went off the air in November 1950. The standard opening in the mid-1940s was low key; Casey and Ethelbert would begin an innocuous conversation which would immediately lead to an Anchor Hocking commercial by the announcer Tony Marvin.

The series was brought back to the air in January 1954 as a replacement to fill the time slot for *On Stage,* with Elliott and Cathy Lewis, on Thursday nights for CBS, while that couple went on an extended vacation. After a few weeks, when *On Stage* returned, CBS kept both series on the air, and relegated Elliott and Cathy's show to follow the re-titled *Crime Photographer*. The latter remained on the air for another sixteen months before being terminated in April 1955. The new 50's opening went like this:

ANNOUNCER: First on the scene: Crime Photographer!

SOUND: (*Flash bulb and shutter click*)

CASEY: Got it! Look for it in *The Morning Express*.

SOUND: (*Organ sting, dissolving into theme*)

In addition to all the name changes, CBS also varied its broadcast times and days of the week. Over that twelve years, the show's starting time was anywhere from 9 PM to 11:30 PM, and its rotation schedule had touched every evening of the week, except Sundays.

When the show debuted in the summer of 1943, Matt Crowley played Casey for a few weeks; he was followed by Jim Backus. By late 1943, Cotsworth had secured the role and he never relinquished it for the rest of the run. Born in 1908 in Oak Park, Illinois, Cotsworth, as did most of his radio contemporaries, began on the stage. In fact, he met his wife, Muriel Kirkland, when they were both in the cast of Noel Coward's *Private Lives*; she later became a radio performer too. The steady pay, flexibility, and increasing opportunities in radio lured both of them from the stage to the microphone. Cotsworth was on several soap operas (*Big Sister, Amanda of Honeymoon Hill,* etc.) in addition to adventure drama series, such as *Cavalcade of America, Mark Trail,* and *Mr. and Mrs. North,* in which he briefly played Bill Weigand. As Casey, he became accustomed to seeing the woman playing his co-lead change, nearly every other year.

Jone Allison played Ann Williams for the first year, and Alice Reinheart portrayed her the second year. About 1945, Lesley Woods became the new voice of Ann, and she remained at the microphone with Cotsworth until approximately March 1947, when she left the series. Jan Miner was given the role in the spring of 1947, and Betty Furness may have played the part between Woods' departure and Miner's arrival. Thereafter, except for a few episodes, Miner portrayed Ann Williams for the rest of the time the series was on the air.

Most of Jone Allison's radio career was spent in the casts of various soap operas, *Brave Tomorrow* , *The Guiding Light, When a Girl Marries,* etc. She got occasional comedy roles, i.e. she briefly played the sister of *Henry Aldrich,* and was in the cast of the crime drama, *Cloak and Dagger.* Her role as Ann Williams is her only documented co-lead in a radio series.

Alice Reinheart, a native of San Francisco, born in 1910, could, and did play major roles in every type of network drama. She had leading parts in several women's daily serials, *Life Can Be Beautiful, My True Story,* and *Romance,* etc. as well as anthology dramas, *Silver Theater, Inner Sanctum Mysteries,* and *Grand Central Station.* In addition to playing Ann Williams, Rineheart was also the female lead in another detective series, *Abbott Mysteries.*

An attractive blonde who could have doubled for Veronica Lake in the movies, Lesley Woods found her acting niche first in radio, where only her friends and co-employees knew of her beauty. Born and raised in Iowa, she studied theater and dance; her stage acting, including a few Broadway shows, led her to the microphone where she stayed until television overthrew radio drama. Woods had a voice that sounded strong and savvy and she specialized in female leads in crime shows. She was in the supporting cast of *This is Your FBI* and *Inner Sanctum Mysteries,* and in addition to *Casey, Crime Photographer,* she was, at different times, also the co-lead in *The Shadow* and *Boston Blackie.*

Jan Miner grew up in Massachusetts, studied stage design, but quickly found jobs that were more plentiful in radio. She began her microphone career at WEEI in Boston, and after two years there, successfully made the transition to network programs in New York City. Miner did well in leading roles in *Hilltop House* and *The Second Mrs. Burton,* but her confident voice was better used in adventure series. At various times, she was the co-lead in *Casey, Crime Photographer, Perry Mason,* and *Boston Blackie.*

The premise of the long-running *Casey, Crime Photographer* (under all of its titles) evolved from the adventures of Jack Casey, a staff photographer on a mythical newspaper, *The Morning Express,* and his co-employee, Ann Williams, a reporter. There were two other regular characters on the show, Captain Bill Logan, a homicide detective played by Bernard Lenrow, and the bartender of the Blue Note Cafe, Ethelbert. Casey's first name was traditionally never mentioned on the show, nor was Ethelbert's last name. The man behind the bar was portrayed by John Gibson, who made a career as the voice of "third wheels" on crime adventure shows. (See additional details on Gibson under Kitty Piper in chapter 3.)

Vintage radio historian John Dunning very accurately characterized *Casey, Crime Photographer* as "more history than substance. It was a B-grade radio detective show, on a par perhaps with *The Falcon...*" Listeners today can confirm Dunning's assessment through any of the 72 surviving audio copies of the 435 episodes that aired. Casey and Ann, both full-time newspaper employees, never went to their office; when they weren't solving a crime, they were chatting with Ethelbert in The Blue Note or swinging by the Homicide Division to find material for the next day's scoop. In this era before cell phones, their newsroom supervisors could only contact them by telephoning the Blue Note or Homicide.

Romance frequently distracted crime-solving couples on radio, but any such sentiment between Casey and Ann was strictly on the back burner. This twosome had a cordial, comfortable business relationship, but their familiarity did not breed contempt or courtship. For the first several years, terms of endearment were rarely exchanged, but in the late 1940s, an infrequent "Honey" would be inserted in their respective dialogue. Sometimes their conversation and actions resembled a playground squabble; in a July 1947 episode, when Casey declined Ann's invitation to a moonlight stroll, she kicked him in the shin.

Most, but not all, of their mysteries revolved around a homicide. Customarily, the first lead came from Logan, and they would accompany him to the crime scene, where the three examined clues and questioned witnesses. Next, Casey and Ann would go to the Blue Note to discuss their theory of the case with the noncommittal Ethelbert, while piano music filled the background. Later, Casey and/or Ann would come up with a solution and they would join Logan in arresting the culprit.

Ann's involvement in these adventures ranged from negligible to total immersion, depending on the script. Infrequently, she was restricted to sitting on a barstool at the Blue Note while Casey went out and cracked the case. More often, they went to a crime scene together, gathered evidence with Logan, interrogated suspects, and then, occasionally, went off on their own and blundered into a killer's trap. In the episode of December 19, 1946 entitled "Christmas Shopping," Casey and Ann traced some ransom money to a crooks' hideout in a nearby village. But, in the course of their inquires there, they were kidnapped by two criminals who bound their mouths, wrists, and ankles with tape, stuffed them in an automobile in a locked garage, and then the crooks started the vehicle's engine so that the resulting fumes would permanently end the careers of the two *Morning Express* employees.

In several episodes, Ann Williams proved her bravery, intellectual prowess, and acting talent when Logan or Casey gave her an undercover assignment. "The Chivalrous Gunman" episode of August 14, 1947 had Ann not only pretend to befriend a gangster's moll, she also convinced the woman to testify against her trigger-happy boyfriend, thus putting the thug behind bars.

Another successful undercover operation handled by Ann was the climax of the January 13, 1954 adventure, "Road Angel," which began with a series of robberies and murders on the highway by an elusive female hitchhiker. Casey drove the same highway, eventually found and picked up this hitchhiker, who promptly pulled a gun on him and after a few minutes, the intrepid photographer was left behind on the side of the road, watching her drive away with his car...and his wallet. Ann later retraced his route in her own automobile, with a tape recorder hidden in expensive luggage in the backseat. When she picked up the hitchhiker, Ann assumed the guise of a big-time con-artist and she characterized the hitchhiker's criminal enterprises as small stuff. She got the hitchhiker to brag about her recent crime wave, all of which was captured on tape. Impressed with Ann's car and luggage, the hitchhiker drew her gun on the reporter, but Ann disarmed her, and turned her over to Logan and Casey who had been following in an unmarked car.

In most series featuring a couple who solved crimes together, scriptwriters usually had in reserve an episode in which one of the co-leads did not appear. This was a reasonable precaution so in the event of a sudden illness or injury to one of the leads, the remaining one could carry on alone. Jan Miner got her chance on August 21, 1947 in the mystery adventure "Busman's Holiday." With Casey "out of town," Ann Williams made the acquaintance of a teenaged waitress, whose boyfriend had been recently arrested for a jewelry theft, of which he was not guilty. With a little help from Logan, Ann adroitly unmasked the real thief, discovered his hiding place for the stolen loot, and with his arrest, exonerated the young boy friend of the waitress.

Casey, Crime Photographer had but limited success in other media. In August 1949, *Marvel Comics* launched a comic book series featuring Casey as the central character, with Vern Henkel doing the art work. Staats Cotsworth's photograph was used on the covers, but he had to subject himself to some corny poses. On the December 1949 cover illustration, he had his feet on his desk, next to his press camera, while he smoked a cigarette and read a newspaper. Tugging on his leg was Ann Williams (or a model impersonating her) who tried to direct his attention to a nearby window where an irate thug brandished an automatic pistol. This comic book series attracted few customers; it was canceled in February 1950 after only four issues.

Television was not any more receptive to Casey's charms or detective skills. CBS brought *Crime Photographer* to the small screen in April 1951, but it barely lasted one season. Richard Carlyle was the new Casey, while John Gibson got to repeat his role as Ethelbert. These two were replaced after only three months and their places were taken by, respectively, Darren McGavin and Cliff Hall. These two remained in their roles until the show was canceled in June 1952. Jan Miner was the only other radio cast member who got to play her regular role on the television series. When McGavin and Hall joined the cast, CBS also added another companion for Casey; a cub reporter, Jack Lipman, portrayed by Archie Smith, who wrote the copy to accompany Casey's pictures.

After they left the *Casey, Crime Photographer* radio series, most of the leads remained gainfully employed in their chosen profession, alternately finding work in television, motion pictures and the stage. Reinheart appeared in a number of films, including *The Lieutenant Wore Skirts* (1956) and *A House is Not a Home* (1957) in addition to being seen regularly on television programs, primarily *I Dream of Jeanie, Make Room for Daddy,* and *The Donna Reed Show.* Reinheart died in June 1993 at the age of 83.

Cotsworth also found jobs in both the television and film industry. He was on the silver screen in *Peyton Place* (1957) and *They Might Be Giants* (1971); he also appeared in over a dozen television series. His first was *Studio One* in 1954 and his last one was *Bonanza* in 1972. His engaging voice was silenced in April 1979; he was 71 years old.

Jone Allison retained the role of Meta Bauer on *The Guiding Light* for almost four years (1949 to 1952) thus becoming the last woman to play the part on radio and the first to portray her on television. She also had a major role on another TV soap opera, *As the World Turns,* and was briefly a dancer on *The Ed Sullivan Show* .

Active on both the stage and television after relocating to Hollywood, Lesley Woods won roles on several television soap operas, *Edge of Night, The Nurses, Bright Promise, General Hospital,* etc. In her later years, she returned to the stage, and actively participated in productions of Theatre West in Los Angeles. Woods appeared on their stage, her last role was in *Tom Tom on a Rooftop* with Philip Abbott, and she also donated the current marquee on that theatre. She had reached 90 years of age when she died in August 2003.

Possibly the most successful of the former cast of *Casey, Crime Photographer* was Jan Miner, who as of this writing (2003) is alive at the age of 86. She has done movies, usually in character parts, such as *The Swimmer* (1968), *Lenny* (1974), *Willie and Phil* (1980) and *Mermaids* (1990.) Miner has also performed on stage and recorded some Books On Tape. Her television roles have spanned the years from *Studio One,* in 1955, to *One Day at a Time,* 1978, to *Law & Order* , 1994, to her most recent, *Remember WENN* in 1997. But she probably made the most money in her acting from the residuals on her long-running television commercials as Madge the Manicurist, extolling the virtues of Palmolive liquid dishwashing detergent.

KAY FIELDS

In 1937 an operative of the U.S. Secret Service made the jump from his comic strip to radio, via a syndicated series, *Dan Dunn, Secret Operative # 48.* As Leonidas Witherall, upon his arrival on network radio, was presented with a lady partner for the first time, so was Dan Dunn. A character never appearing in his many previous adventures, Kay Fields, Secret Operative # 185, became Dunn's co-employee and partner in his radio crime-fighting.

If imitation is truly the most sincere form of flattery, then a Chicago artist, Norman Marsh flattered many of his contemporary comic strip creators. When Marsh began his *Dan Dunn* comic strip in October 1933 for Publishers Syndicate, even juvenile readers recognized it as an obvious rip-off of the Tribune's *Dick Tracy,* which Chester Gould originated exactly two years earlier, in October 1931. Dunn had the square jaw, flinty gaze, and snap brim hat, usually drawn in profile, just like Tracy. The only significant difference was their employment; Tracy was a police detective and Dunn was a federal operative.

But Marsh did not confine himself to copying from just one strip. Orphans with dogs were popular in that era; *Little Orphan Annie* with Sandy debuted in 1924 in the Tribune and *Little Annie Rooney* with Zero arrived in 1929 via King Features. So Dunn acquired a large dog named Wolf and then Dunn adopted an orphan named Babs. She had dark curly hair like Annie Rooney and wore a red dress with a white collar like Orphan Annie. Dunn would be missing from his own strip for weeks, during which Babs and Wolf had exciting adventures.

When Marsh decided to give Dunn a funny sidekick, he came up with Irwin Higgs, a portly, slow-witted man wearing a derby. Other than the fact that Higgs had no special fondness for hamburgers, he was almost the mirror image of Wimpy, Popeye's companion in *Thimble Theater.* Marsh even borrowed arch-villains; his Wu Fang was just another Fu Manchu. Why the Tribune, or other syndicates, did not pick up the telephone and contact *Superman's* attorneys (who would later eradicate *Captain Marvel)* and sic them on Marsh has never been determined.

Despite, or perhaps because of, his literary piracy, Marsh turned Dunn into a fairly popular fictional hero. His strip made it to comic book status in February 1935, a full year before *Dick Tracy* did. Dunn also appeared in two pulp magazines in 1936 and his comic strip fame resulted in his exploits being published in seven Big Little Books (from 1934 to 1940) which made him one of the front runners in that juvenile market.

The syndicated radio series, *Dan Dunn, Secret Operative # 48,* was probably first released in 1937 and only four audio copies of these 15-minute episodes have survived. The title lead was played by Lou Marcelle, who was also the voice of radio's *Fu Manchu;* he supplied the opening narration for *Casablanca,* among other films. Since the only known female in the cast was Lucille Meredith, we may assume she was the voice of Kay Fields. Other supporting players were Myron Gary, David Sterling, Gerald Mohr, and Hans Conried, whose gift for foreign accents made him the most versatile member of the cast.

Radio historian Jim Cox has identified three writers who worked on the scripts: Maurice Zimm and the husband and wife team of Paul and Ruth Henning. That couple were down on their luck when they were offered $15 an episode to write scripts for the first 13 episodes. When they finished, they were paid only half of what they were promised but it was enough money to get them back home for Christmas, so they did not pursue getting the rest they were owed. It's certainly possible that Ruth's contributions made Kay Fields such a strong character.

As with most syndicated shows, each 15-minute episode contained at least three minutes of only music, usually split evenly between the beginning and end of each program. This would enable the local advertiser's commercial copy to be read over the music by a station announcer. In the *Dan Dunn* series, a rousing military march was chosen. More than one radio researcher has cynically commented that the music was the best part of *Dan Dunn's* radio program. After the beginning music ended, the syndication announcer crisply gave the radio listeners the below introduction:

ANNOUNCER: The thrilling exploits of Dan Dunn, Secret
Operative # 48, are known to some 75 million
readers. Now at last, this famous cartoon newspaper character, created by Norman Marsh, is
brought to the air lanes in a sensational series of
new adventures that will enthrall young and old!

Just as his comic strip scenarios had no basis in law enforcement
reality, Dunn's federal assignments on the radio show lacked any factual
background. The U.S. Secret Service had two basic responsibilities: protection of the President and suppression of counterfeiting. But Dunn,
whose motto was "Crime Never Pays," investigated gangsters, railroad
robbers, smugglers, drug dealers, assassins, and assorted spies and saboteurs. Another factitious part of both the comic strip and radio story-line
was the constant presence of Irwin Higgs, described on the radio as "Dan's
old friend and inseparable companion." Higgs, an inexperienced, but apparently well-meaning, civilian accompanied Dan and Kay on all their
official business, which would be contrary to the regulations of every federal law enforcement agency. The only saving grace in radio's Higgs was
that he was no longer the funny stumble bum of the comic strip, and had
matured into a sagacious character.

On their radio show, Dan and Kay were not just Secret Service co-
employees, they were closely-attuned partners who frequently finished
each other's sentences:

KAY: The importance of those blueprints to the military
strength of our country can't be measured in money
alone; isn't that right, Dan?

DAN: Yes, and that's why we've got to overtake Lobarr. If
he gets away, we'll have lost our only—

KAY: We understand that; you don't think because I
cried out back there it was because—

DAN: Because you were afraid? Oh no! You're the bravest
woman I've ever known.

HIGGS: Yeah, and then some!

One of the cases detailed in the surviving audio copies concerned the efforts of the Secret Service to prevent sabotage to an experimental bomber, the Z-19. Dan described this battle in pulp magazine terminology:

DAN: A giant international spy ring of vicious and de-based spies the world has never seen! They sell their services to the highest bidder on the open market!

Although Dan had a tendency to bark orders at Kay, she was certainly not his subordinate. If anything, she appeared to be the brains of the outfit. Several scenes in their radio adventures illustrate this point as Kay tried to rein in Dan, a fellow who foolishly responded to most challenges with total recklessness. In one incident, ignoring Kay's warning, he ran in front of a departing airplane and was almost killed. Later, trying to apprehend a saboteur in a speeding auto, Dan drove through a red light, against Kay's will, and he crashed into a third car, thus sending Kay to the hospital. Finally, without advising Kay or their Chief, Dunn blundered into a trap and was kidnapped by evil-doers. Fortunately, Kay had learned of his intentions, followed him to the kidnap scene, and wisely called the Chief for back-up.

All in all, this radio series was neither logical nor entertaining. Dan Dunn, who appeared in the comic strips, Big Little Books, pulp magazines, and syndicated radio, seemed to illustrate the weaknesses of each different venue. However Kay Fields was probably responsible for the only unique quality of this series since she was the first woman federal investigator portrayed in the history of broadcasting. There would be only one other: Helen Holden, who came to the air in 1941.

A Nose For News...And Clues

The people who created radio characters and wrote the scripts of their adventures tended to place their characters in occupations which would provide fertile ground for story-lines. For example, in the soap operas, there were a large number of characters in the practice of medicine and the ministry. Juvenile serials frequently involved the activities of an aviator, a cowboy, or both in the case of *Sky King*. Crime dramas routinely put their heroes and heroines into the employment of a news medium since each assignment from their editor brought with it a new crime to solve.

WENDY WARREN

A CBS soap opera with inclinations toward detective drama, *Wendy Warren and the News* first came to the airwaves at noon on June 23, 1947. Radio historian John Dunning is convinced the series was primarily a clever trap set by the network: "...to snare listeners who might otherwise not be caught dead listening to a soap opera. Scheduled in the newsy time slot of high noon, it opened each day with three minutes of straight news read by...CBS newsman Douglas Edwards."

Following these actual news bulletins, Edwards turned the microphone over to Wendy Warren (the voice of Florence Freeman) who then read about a minute of real news of interest to women. When Warren finished, the "news program" concluded with a Maxwell House Coffee commercial. Next listeners heard Edwards and Warren, supposedly "off-mike," exchange some brief, personal (but scripted) remarks which would be a seamless segue into Warren's daily drama, some of which began with a phone call to her mythical radio studio.

Florence Freeman, *Wendy Warren and the News* (*The Stumpf-Ohmart Collection*)

Whether it was truly a clever trap or not, it certainly was a unique concept for a successful radio program. The series ran for over a decade, probably broadcasting about 2,800 episodes, of which only 16 audio copies have survived. Based on those extant episodes, the following average time line was utilized:

00:00 to 00:15	Warren introduces Edwards
00:15 to 02:15	News bulletins by Edwards
02:15 to 03:05	News bulletins by Warren
03:05 to 04:05	Maxwell House Coffee commercial
04:05 to 04:20	Warren and Edwards chat "off-mike"
04:20 to 12:00	Dramatization of Wendy's Adventures
12:00 to 13:15	Baker's Coconut commercial
13:15 to 14:00	Announcer previews next episode
14:00 to 15:00	Second Maxwell commercial and sign-off

Since the dramatic portion of this quarter-hour show only consumed about seven and a half minutes, churning out this daily script must have been relatively easy for writers John Picard and Frank Provo. Over the years, they alternated the standard soap opera themes of unrequited love, matrimonial strife, and a woman's quest for happiness with Wendy's crime-solving endeavors. Incidentally, her name was always Warren, but she was also Mrs. Gilbert Kendal and (almost) Mrs. Mark Douglas. But she probably retained her professional name on her radio news program so as not to confuse her audience. Such a name change would also affect her by-line at *The Manhattan Gazette,* for the hard-working Warren not only had a daily radio news program, she also was a correspondent for that mythical New York City daily newspaper.

The most comprehensive analysis of *Wendy Warren and the News* is contained in a separate chapter in Jim Cox's *The Great Radio Soap Operas,* which was published by McFarland & Company in 1999. Cox explained that CBS created this series to fill their time slot vacancy created when *The Kate Smith Show* was moved by the sponsor to the Mutual Network. Cox concluded, "Radio produced a hybrid show between fact and fiction—often making the part that was never intended to be believed sound as if, in reality, it could be accepted too."

The three mainstays of the series were Florence Freeman (playing Wendy), announcer Hugh James (playing announcer Bill Flood) and Douglas Edwards (playing himself). Freeman, a Manhattan native with a Master's Degree from Columbia University, began her radio career in her mid-20's and went on to become one of the most successful performers in women's daily serials. Most of her network contemporaries would have been delighted to have the lead in just one major daily serial, but Freeman had the lead in about a half-dozen programs, including two at the same time. At different periods, she was the female lead in *Dot and Will, The Open Door, Valiant Lady* and *A Woman of America.* During most of the time she was starring in *Wendy Warren and the News,* she also had the title role in *Young Widder Brown.*

Hugh James, born in the Bronx in 1915, followed the traditional path of many network announcers by beginning as a page boy at NBC studios, advancing to tour guide, and then as an apprentice announcer limited to giving station breaks. After customary stints at the NBC affiliates in Philadelphia and Washington, DC, James returned to NBC headquarters in New York City where he eventually was selected as the primary announcer on approximately a dozen series over the next twenty years. While many of his assignments were soap operas, he was also on crime dramas (*True Detective Mysteries*) and musical anthologies (*The Voice of Firestone*).

Born in Oklahoma, Edwards was two years younger than Hugh James. He spent his childhood in New Mexico and was fascinated with what he could hear on his crystal radio set. Before he got out of high school in Troy, Alabama, Edwards had decided on his life's work. Throughout high school and college, he worked temporarily at small local radio stations in Alabama and Georgia, before landing his first full time microphone job at the heralded radio station, WXYZ in Detroit. Here Edwards had a program where he was called "The Cunningham Ace," named for his sponsor, Cunningham's Drug Store. He also shared staff announcer duties with Mike Wallace, and others, on *The Lone Ranger* and *The Green Hornet.*

Edwards made the big jump to CBS News in 1942, but had to serve a year of being a "news reader" before he took over the CBS microphone from John Daly, who had been transferred to North Africa to cover the war there. Despite his pleas for overseas duty, Edwards remained in the U.S. until, finally, in the closing months of World War II, Edward R. Murrow brought him to France. CBS kept Edwards in Paris as a correspondent to prepare for the Nuremberg trials, so he did not return to the States until June 1946. Shortly after, CBS president Frank Stanton picked Edwards to head up the network's television news.

At first, Edwards fought the assignment as he knew it meant more work for less money; he would have to forego the lucrative income that radio advertising could provide. The technical problems were large, particularly how to keep viewers interested in a man looking down, reading a news bulletin. Not until the invention of a rudimentary teleprompter was this problem solved. So, by the time Douglas Edwards began his stint on *Wendy Warren and the News*, he was well known to every family on the East Coast, who had a television set, from his nightly CBS show, *Douglas Edwards and the News.*

For the first two years, most of Wendy's adventures were fairly standard soap opera fare. Her childhood sweetheart and on-again, off-again fiancée, Mark Douglas, was the source of either joy, anger, confusion, sadness, or hope for Wendy, depending on the story-line. She eventually married megabucks publisher and thorough cad, Gilbert Kendal, who was pursued by two seductresses, Nona Marsh and Adele Lang. Nona was once married to Mark Douglas, while Adele's husband, Charles, an unscrupulous land dealer, eventually had her committed to a sanitarium with bars on the windows. It may have been difficult to tell the players without a scorecard.

In one story-line in the spring of 1948, Kendal, then married to Wendy, was lured to the apartment of Nona, where he complained that his wife wanted to move out to Long Island, which would make his commute long and tiresome. (Even then, the New York region had commuter problems.) Kendal and Nona speculated on the whereabouts of Adele, whose husband was trying to swindle her. About a week later, listeners heard Wendy formulate a clever ruse to uncover the secret location where Adele was being held hostage in a sanitarium. Wendy instructed her maid, Bertha, to ingratiate herself with Mr. Lang's housekeeper so Bertha could discover Adele's location and then report back to Wendy.

By 1949, Wendy's adventures had taken on the trappings of a detective drama, or more accurately, those of a counterspy. Warren, working with a private investigator, Rusty Doyle, and her compatriot, Anton Kamp (played with a thick European accent by Peter Capell), strove to uncover a massive conspiracy against the free world. As part of her plan, Wendy used some undercover techniques to penetrate "The Charmed Circle" but it almost backfired. A woman of intrigue, Madeline Marstow, had her brutish bodyguard, Hugo (who also had a thick European accent), drag Wendy to Madeline's high rise apartment. She then threatened Wendy of dire consequences unless our heroine obtained data on "the Murdock case" from Rusty Doyle, which Wendy then had to convey to Madeline.

Just how well Wendy fared in her battles against the forces of international crime, we will leave to future radio historians. It stands to reason that it took more courage, intellect, and investigative skills to thwart global evildoers than solving the routine crimes of robbery, kidnapping and murder that other lady detectives faced. With the possible exception of Diane LaVolta (see chapter 2), Wendy Warren was the only woman crimefighter who battled international criminals.

By the time that *Wendy Warren and the News* finally ended in November 1958, Douglas Edwards was in the forefront of American television newscasters, anchoring the *CBS News* broadcasts. He also hosted *Armstrong Circle Theater, Masquerade Party,* and was the anchorman for the Miss America pageant. Many honors were garnered by Edwards, including Emmy Awards in 1956, 1958, and 1961, plus the Peabody Award for his coverage of the *Andrea Doria* sinking. However, in April 1962, the competent but bland Edwards was replaced by a more energetic Walter Cronkite as the *CBS News* anchor. Edwards, though deeply disappointed, was the first one to congratulate Cronkite, who later said it was "the classiest damn thing I ever saw." Despite the demotion, Edwards stayed on at CBS, doing both television and radio in a reduced role, for another 26 years before retiring in 1988 at the age of 71. He had cancer and would battle it for another two years in Florida before dying on October 13, 1990.

Freeman, after the end of dramatic radio, did not need employment in other media. She was the wife of a rabbi so her family and her religion were very important to her. After her daily shows on radio ended, she could devote more time to her personal and religious life, and she did. Freeman and her husband retired to south Florida , where she loved to play bridge with friends, including Vivian Smolen Klein, a close associate of hers from the soap opera days. Vivian was the heroine in *Our Gal Sunday* and also played Lolly-Baby on *Stella Dallas.* In 1998, Freeman came to the Friends of Old Time Radio convention in Newark and was delighted to greet her former co-employees, including George Ansbro, her long-time announcer on *Young Widder Brown,* as well as many of her new fans. Her husband died in 1999 and she moved to New Jersey where she lived with one of her daughters. Later she moved to northern Illinois to live with another daughter. In the spring of 2001, this lovely lady died at the age of 90.

ANNE ROGERS

Hot Copy was not, as its title may have suggested, a story of a sleazy tabloid, rather it was the adventures of a successful, crime-solving reporter, Anne Rogers, who worked for a responsible daily. NBC never could decide what was the best time to air this series about a feminine sleuth. When it first reached the airwaves in October 1941, this half-hour program was scheduled for 10:30 p.m. on Saturdays. Then NBC moved it to Mondays at 11:30 p.m. for the summer of 1942, however in the fall, the network changed it to 9:30 p.m. on Saturdays. In July 1943 the series was

Betty Lou Gerson, *Hot Copy*

time-shifted again to be broadcast at 3:30 p.m. on Sundays. When it went off the air in November 1944, NBC was airing it at 5:30 on Sunday afternoons. Albert Crews directed the entire run of approximately three years.

Radio historian Jim Cox has determined that the heroine was originally named Patricia Murphy, but later acquired the name of Anne Rogers. For most of the run, the standard opening was rather unpretentious; the announcer would open with:

"O-Cedar, the greatest name in housecleaning, presents…Hot Copy!"

This was immediately punctuated by an organ sting, and then a scene of about 15 seconds followed which detailed the evening's crime. The announcer then introduced the drama with these words:

"And now…a new and dramatic story of Anne Rogers' search for Hot Copy!"

While NBC's shuffling of the program's schedule had little to do with the leading roles, there were three women who, in succession, played the part of Anne Rogers; they were Betty Lou Gerson, Eloise Kummer, and Fern Persons. It would be difficult to name a radio series in which Gerson never played a part. Originally from Tennessee, she began on radio when she was twenty, and by the time she was forty, her versatility and superb acting had taken her through hundreds of radio roles. Gerson was heard on *Midstream, Don Winslow of the Navy, The Woman in White, The Adventures of Philip Marlowe, The Guiding Light, The Whistler, Today's Children, Inner Sanctum Mysteries, Attorney at Law,* and *Mr. First Nighter,* as well as many others.

Kummer spent most of her broadcasting career toiling in the soap operas, playing parts on *Lone Journey, Backstage Wife, The Right to Happiness, The Story of Mary Marlin,* and *The Guiding Light.* Persons had less experience than the other two women, but she had been in the supporting cast of *Midstream* when Gerson was the co-lead. *The Bartons, Author's Playhouse,* and *Helpmate* were other shows where Persons found work.

Portraying the part of Sergeant Flannigan, the associate and foil of Rogers, was Hugh Rowlands. He had begun in Chicago radio as a child actor, who after his voice changed, could still play little kid parts by pitching his voice higher. He did this on *The Tom Mix Show,* playing a juvenile, Jimmie, in the mid-1930s when he was in his early twenties.

In *Hot Copy,* Anne Rogers used her newspaper column, "Second Glance," as a springboard to investigate murder, robbery, and other assorted crimes. Spritely Poole, a none-too-bright co-employee of Anne's, was also her roommate. but most of the investigative phase of any case was shared by Anne with a local policeman, Sergeant Flannigan, a well-meaning, but mediocre detective.

Betty Lou Gerson of *Hot Copy* (*The Stumpf-Ohmart Collection*)

Nelson S. Bond, the scriptwriter for the series, had a knack for twisting an old expression into an apt phrase humorously. When Rogers and Flannigan arrived together at a murder crime scene, the following exchange took place:

FLANNIGAN: This is the place. Now, remember what I

said, Miss Rogers. I like a pretty woman as well as the next guy, but when there's a job to be done, a woman...

ROGERS: Should be scenery, not heard. Very well, Sergeant, I understand.

This odd couple had a cooperative relationship, and despite their differing approaches to each crime, managed to unravel every mystery, with Rogers always leading the way. Flannigan occasionally called her "Miss Sherlock," as much in jest as in admiration. At the conclusion of each episode, although Rogers had almost single-handedly solved the mystery, her police officer chum claimed that "we" had cracked another case. Never one to crush his illusions, Rogers merely chuckled.

From all indications, this was an excellent series, with crisp plotting, engaging mysteries, and generous humor. The bad news is that not one audio copy has yet been found. The good news is that one complete script has been located and perhaps more will be in the future. The script is undated, but is from the 1944 era, and is entitled: "Death to Play and Mate."

The episode begins when Rogers and Flannigan are alerted to the murder of Dr. Warren King, a retired physician, at his residence. His corpse was found near a chessboard in his study. The suspects are his adopted daughter, her boyfriend, and Lincoln Owens and John Bishop, two men who had recently quarreled with the doctor. After Flannigan and Rogers examine the crime scene and question all the suspects, she correctly deduces that Bishop is the killer. After he is led away in handcuffs, Flannigan drives Rogers home to her apartment and the following conversation takes place in his patrol car. Note how gracious Anne is in upgrading Flannigan's meager contribution to the solution of the crime, generously using 'we" in place of "I."

ROGERS: As we now know, when Bishop broke in on Doctor King, the old man knew his moments were numbered.

FLANNIGAN: Yeah, *we* know that.

ROGERS: We can assume that King tried to argue him out of murder...

FLANNIGAN: Yeah, so *we* can.

ROGERS: But meanwhile, King arranged the pieces to reveal
            Bishop's name and intention. Since Bishop didn't
            play chess, he didn't know what was being done.
            So…we got the message.

FLANNIGAN: Yeah. So we did.

ROGERS: But, of course the evidence was too flimsy to convict
            a man in court. So we had to break Bishop with a
            sudden accusation. And we did…and it worked.

FLANNIGAN: Uh-huh, pretty good! We done all right, didn't
            we? Well, Miss Rogers, now you can see us cops
            crack cases like this. You come around some time
            again…and if there's ever anything you want to
            know…just call on me!

ROGERS: (*Laughing*) Oh, Sergeant Flannigan!

Not a great deal is known of the subsequent broadcast careers of
Kummer and Persons, but Gerson went on to roles in television and also
did vocalizations for animated films. She was featured on *Morning Star*, a
mid-sixties television series, and she provided the evil voice of Cruella
DeVil in Disney's full -length cartoon, *101 Dalmatians*. Hugh Rowlands
was 64 when he died in January 1978.

CALAMITY JANE

The series *Calamity Jane* was the only one in the history of American
broadcasting that was created solely to fill the vacancy brought about by
the death of a Caucasian male actor who was portraying an African-American female lead.

The Beulah Show was a spin-off from the radio comedy, *Fibber McGee
and Molly*. Marlin Hurt, a white man, was the voice of Beulah, the Negro
maid of the McGees', and this character became so popular, "she" got her
own show. *The Beulah Show* debuted on CBS in July 1945 and had enjoyed excellent ratings when suddenly, in March 1946, Hurt suffered a
fatal heart attack. He was only in his mid-40s at the time.

The network scrambled for a replacement series and Agnes
Moorehead agreed to take the title role in a new show, *The Amazing Mrs.
Danberry*, which Helen Mack would direct, just as she had done with *The
Beulah Show*. But it would take about three weeks to complete the scripts

and other production details for *The Amazing Mrs. Danberry* so another series was hastily thrown together in one week for Moorehead. This temporary program was called *Calamity Jane.*

The entire run of *Calamity Jane* consisted of just three weekly performances, from March 30 to April 14, 1946, and in it Moorehead portrayed a zany reporter who uncovered criminal enterprises in her community, bringing the light of justice to gangsters and con artists. Since the show aired only three times, it is to be expected that no audio copies nor scripts have been located; moreover the brevity of the show's run did not permit radio historians to catch her last name. We do know that three other performers were in this series who, like Moorehead, were marking time before taking over their respective roles in *The Amazing Mrs. Danberry*: Bill Johnstone, Cathy Lewis, and Don Wolfe. The latter was the voice of Jane's grandfather, and the publisher of their newspaper, in *Calamity Jane.*

But *The Amazing Mrs. Danberry,* despite the talent around the microphone, also sank soon with hardly a trace. Moorehead, in this series, was a rambunctious widow, whose meddling in her son's attempts to manage their department store, seemed to be a good basis for humor. But the series never caught on with the radio listenership and it was canceled in June 1946, after only two months.

These two quick cancellations certainly had no effect on the show business career of Moorehead, one of the most talented performers on radio, stage, silver screen, and much later, television. She was born December 6, 1906 in Clinton, Massachusetts and made her professional debut at age 11 in the chorus of the St. Louis "Muny" Opera, an outdoor summer musical theater. After graduating from college at the University of Wisconsin, she briefly taught speech and drama in high school, while gaining experience in regional theater. By 1930, she had relocated to New York City and was finding some work in radio, together with small parts on Broadway. Moorehead next toured in vaudeville with Phil Baker for three years, ending in 1936.

Orson Welles hired her for his Mercury Theatre Company in 1940, which provided her with major roles in both his radio and stage productions. He also cast her in his 1941 classic film, *Citizen Kane* and the next year, she was nominated for an Academy Award for Best Supporting Actress in his motion picture, *The Magnificent Ambersons.* On network radio in the 1940s, she played every type of role, from preschool girls to querulous old ladies, and was heard on *The March of Time, Cavalcade of America, Mayor of the Town, The Aldrich Family, Terry and the Pirates,* and *Brenda Curtis.*

Between radio assignments, her talents were used by Hollywood in *Jane Eyre* (1943), *Tomorrow the World* (1944), *Dark Passage* (1947) and others. In the 1950s she did less radio work, concentrating on film projects, as well as touring with Charles Laughton and Charles Boyer, doing staged readings of George Bernard Shaw's *Don Juan in Hell*. The next generation of fans were delighted with her television role of "Endora" on ABC's *Bewitched* in the late 1960s, and moviegoers saw her in *How the West Was Won* (1962) and *Hush...Hush Sweet Charlotte* (1964).

Moorehead returned to dramatic radio in the early 1970s, shortly before her death, when Himan Brown cast her in episodes of *CBS Radio Mystery Theater*. She was 67 when she died in April 1974. Of all the significant roles she had played on stage, in movies, and on radio, probably the one most recalled by her fans today is that of the bedridden wife about to be murdered in the chilling classic by Lucille Fletcher, "Sorry, Wrong Number." It was initially presented on *Suspense* and was so compelling, Moorehead repeated that radio role seven times in subsequent years.

SANDRA MARTIN

Lady of the Press: Sandra Martin was originally called *The Story of Sandra Martin* when it debuted on CBS in May 1944. The series bore the second title from June 1944 to mid-1945 when it went off the air. This series was unique in that the network aired two separate versions with different leads; Janet Waldo played Sandra in a 15-minute show which was broadcast five times a week while Mary Jane Croft did the honors in a half-hour program once a week. However, some sources cite Waldo in the weekly show and Croft in the daily one. Both versions told of the adventures of a crime-busting reporter, and co-publisher, of *The Daily Courier*, who also struggled to find true love.

Miles Laboratories sponsored both versions, on the quarter-hour show they usually touted the relief of tummy distress by taking Alka-Selzer while the announcer was praising the wonderful properties of Bactine antiseptic and One-A-Day vitamins during the thirty-minute program. The 15-minute daily show was a serial, so Sandra might take up to several weeks to solve the mystery and turn the guilty scoundrels over to the local police. While no audio copies or original scripts have yet been uncovered for the half-hour weekly program, it's a safe bet that each of her cases were wrapped up in the allotted thirty minutes.

Mary Jane Croft, a "Lady of the Press" in *Story of Sandra Martin*
(*The Stumpf-Ohmart Collection*)

Both ladies in the leads were shy about giving their true age, but most radio historians speculate that the two of them were born in 1918; Croft was a native of Indiana while Waldo came from the state of Washington. Both specialized in character roles in West coast network shows. Croft was in the supporting cast of *Blondie, The Adventures of Sam Spade, The Beulah Show,* and *The Mel Blanc Show.* Waldo had roles in *The Railroad Hour, The Adventures of Ozzie and Harriet,* and *Silver Theater.* In addition, both ladies appeared in *One Man's Family.* Waldo was well known for portraying the title lead in *Meet Corliss Archer,* an exuberant teenager that she played until she was in her mid-30s.

Sandra Martin was a Los Angeles reporter who proved her dedication, intuition, and occasionally foolhardiness, in her battle against crime. She was sometimes assisted by her love interest, Lt. Hack Taggart (the voice of Ivan Green), who was bitter about his military experiences in World War II. Hack was also jealous about his relationship with her newspaper partner, a photographer named Skip Williams. Gordon Hughes produced and directed both versions of this series; his writers were Leslie Edgely and Dwight Hauser. The supporting cast for both versions included Griff Barnett, Howard McNear, Howard Culver, Eddie Marr, Jay Novello, and Bob Latting. When the weekday program debuted, *Daily Variety* reviewed it favorably:

> "Daytime serials and whodunits not lacking for audience among the housewives, this one should cut its own groove. In the role of the sleuthing sobbie is Janet Waldo, one of the busiest young ladies in radio and well equipped for the dramatics that will keep her hopping."

Lady of the Press was one of those soap opera-detective series, a hybrid similar to *Perry Mason* and *Kitty Keene, Inc.* These series posed challenges for their respective scriptwriters, who had to combine mystery, unrequited love, criminal violence, jealousy, danger, and romance…sometimes all in one 15-minute program. It is not known which of the two writers, Edgely or Hauser, penned the below dialogue from the one surviving audio copy of the 15-minute version, but it demonstrates a typical episode. At the beginning of this segment, Skip burst into a room where Sandra was being held hostage, but her armed captor fled, without harming either newspaper employee. The next day, Sandra and Hack discussed the pain of their relationship problems, complete with pregnant pauses and soft organ music:

HACK: Don't be so sorry for yourself.

SANDRA: I'm not sorry for myself, Hack (pause) I'm sorry for us.

HACK: (Pause) Us?

SANDRA: Yes (pause) for us (pause) and for the things we've lost. (pause) There's an acute shortage in the world of the kind of love and beauty we once had.

	(pause) There isn't enough to waste it…the way we're wasting it.
HACK:	(Pause) Beauty isn't very high on the world market at present.
SANDRA:	All the more reason (pause) that we should have held on to what we had.
HACK:	Sandra (pause) you should learn to face things the way they are (pause) not the way you'd like them to be.

In the concluding scene of this episode, later that evening, Sandra and Skip, at her insistence, had gone to the estate of a suspect and rang his doorbell. As they waited for someone to answer it, an automobile suddenly roared past them, and an unknown gunman fired a hail of bullets at the two. Before the radio audience caught their breath, announcer Dick Cutting was at the microphone to tell them that Alka-Selzer would speedily cure that abdominal too-full feeling from their late night snack.

Barry Weiss, who recently reviewed several of the scripts archived at Thousand Oaks Library near Los Angeles, determined that many of the early scripts involved wartime themes, bordering on propaganda. This included black-market gasoline abuses, counterfeiting of ration cards, and other violations of the regulations of the Office of Price Administration (OPA). In one episode, Hack lamented that there were no strong criminal penalties for OPA violations.

However, by 1945, the story-lines were less concerned with criminal activities and mystery-solving and more with standard soap opera themes of romance, lost love, and assorted heartaches. Sandra was even married to, and later divorced from, Curt Kavanagh, a faithless playboy, posing as a songwriter. After their divorce, he moved back East, only to return professing his deep love for her. Her resultant confusion was resented by Hack, who still loved her. Hack discovered information about Curt's unsavory activities in the East and told Sandra. In the last episode of the series, April 27, 1945, Sandra rejected both men and she felt liberated. "On that note, we close the Adventures of Sandra Martin" were the announcer's last words.

Both Croft and Waldo, after the demise of dramatic radio, found a new generation of fans on television and voiceovers for animated film. Croft became popular as the sexy voice of Cleo, the dog, on television's

People's Choice. Waldo was the voice of "Judy" on *The Jetsons,* was Fred's mother-in-law on *The Flintstones,* and the lead in *Josie and the Pussycats.* Now in her 80s, Waldo is still active in voiceovers and makes personal appearances at nostalgia conventions.

LINDA WEBSTER

Between commercials for Palmolive soap and Colgate brushless shaving cream, newspaper reporters Linda Webster and Jack Winters went behind the headlines to solve major crimes on the CBS radio series, *City Desk.* The network began this program in January 1941 on Thursday evenings. Six months later, it was time-shifted to Saturday nights, where it remained until its demise in September 1941.

Gertrude Warner (1917-1986) had the role of Linda Webster for the entire nine months that the series was broadcast, but for some reason, the actor portraying her co-lead kept changing. Three different actors, in succession, were the voice of Jack Winters; they were Donald Briggs, James Meighan, and Chester Stratton. Warner, a native of West Hartford, Connecticut, got her start in radio performing at Station WTIC in Hartford about the time she graduated from high school in 1935. Within four years, she had relocated to New York City and was getting regular work on NBC's soap operas, including *Against the Storm* and *Valiant Lady.* Her vocal skills brought her to the title roles in *Joyce Jordan, M.D.* and *Ellen Randolph.* (For details on her leading roles in detective dramas, see Della Street entry in chapter 6.)

A hard-working actor in the soap operas, Donald Briggs had major roles in *Girl Alone, Portia Faces Life, David Harum* and many more. He was also in the cast of *Death Valley Days* and Edgar Guest's series, plus had the lead role in four different crime dramas, including *Perry Mason,* in which Gertrude Warner, for a period, was playing Della Street. In addition to their close association in several radio series, Fate had them leave this world, almost together. Briggs, at age 75, died on February 3, 1986, exactly eight days after Warner passed away.

The son of theatrical parents and born in Paterson, New Jersey, Chester Stratton began his radio career after briefly attending Rutgers. He went on to substantial success on both network radio and the Broadway stage. Much of his broadcast work was on the women's daily serials *Big Sister, Against the Storm, Pepper Young's Family,* and *Her Honor, Nancy James.* Stratton also had major roles in adventure series (*Cimarron Tavern* and *Wilderness*), plus he had the title role in *Hop Harrigan,* a juvenile aviation series. On the Broadway stage, he was in *White Oaks,* with Ethel Barrymore in the lead, and he was in the cast of *A Connecticut Yankee.* During a World War II tour of *The Barretts of Wimpole Street,* sponsored by the USO, Stratton appeared in a cast headed by the legendary Katharine Cornell. After dramatic radio was supplanted by television, he relocated to the West coast where he worked in both motion pictures and television. This talented performer died in Los Angeles, California in July 1970, just three weeks shy of his 58th birthday.

The vintage radio biographer, Thomas A. DeLong, described Manhattan-born James Meighan (1906-1970) as a " Perennial daytime lead, notably as matinee idol Larry Noble in *Backstage Wife.* Nephew of silent film star, Thomas Meighan, he honed his craft with the Yonkers Stock Co. and in eight O'Neill plays, including *Desire Under the Elms.* Prior to radio, he appeared on Broadway in *My Maryland and Hamlet in Modern Dress* in the 1920s. Meighan played opposite Helen Hayes in her 1936 radio series and took over from Bert Lytell the lead in *Jimmy Valentine.* "

Meighan's other soap operas were *Against the Storm, Dot and Will, Lone Journey, Just Plain Bill, By Kathleen Norris, The Romance of Helen Trent,* and *Marie, the Little French Princess.* With a voice of authority, he also played the leads in the crime-fighting dramas *Special Agent, Flash Gordon,* and *The Falcon.* Meighan passed away in June 1970 in Huntington, NY at the age of 63.

Since all three of the men who took their turn playing Jack Winter were talented veterans, Gertrude Warner certainly had no complaints on her side of the microphone. Moreover, since not one audio copy of their *City Desk* series has yet surfaced, it will be impossible to determine if any one of the three fellows was superior to the other two in the role.

The supporting cast of *City Desk* included Geoffrey Bryant, who played Dan Tobin, the editor, Jimmy McCallion as a character named Caruso, plus Ethel Owens and Karl Swenson. Both Himan Brown and Kenneth MacGregor directed the program at different times and the script duties were shared by Frank Dahn, Stuart Hawkins, and Frank Gould.

For nine months, in this crime adventure drama, Linda Webster and Jack Winters left their newspaper office regularly and used their brains and pluck to confront crime. We may safely assume they uncovered illegal, sinister plots and solved mysteries of criminal enterprise which had baffled the local police. None of these Herculean tasks, apparently, prevented them from returning to the typewriters and meeting their deadlines. No further details regarding their specific cases or accomplishments can be discussed here as radio researchers have discovered neither any audio copies nor any written scripts.

Me and My Gal Friday

The women in this chapter all performed competent, deductive service in assisting their bosses or boyfriends on the trail of mysterious criminals. While their sleuthing responsibilities were similar, the four ladies discussed here had different occupations. One of them was a journalist for a Manhattan daily publication and another was a news reporter for a small town paper; both of them helped solve crimes their boyfriends were investigating. A third was the housekeeper for a New England academician, while the fourth was the executive assistant to a crime-busting attorney.

DELLA STREET

Perry Mason's secretary, Della Street, was far more to him than her title would suggest. She was Mason's closest associate, a professional and personal companion, and, occasionally, his co-conspirator. This bright and brave lady worked diligently at Mason's side through 82 novels, a radio program that lasted a dozen years, two short-lived comic book series, six Warner Brothers movies, and a television run of nine years in primetime. Since each of these categories had a different writer (or a series of writers) the characters of Perry and Della varied significantly in their personality, attitude, behavior, as well as their relationship to each other.

Erle Stanley Gardner, born in Malden, Massachusetts in 1889, created Perry Mason, largely in his own image. Although Gardner was once expelled from college for punching a professor, he eventually completed law school and went on to create a lucrative legal practice. As an attorney, he wrote fiction in his spare time. Based on the success of his first Perry Mason novel in 1933, *The Case of the Velvet Claws,* he gave up his law practice. Thereafter, for the next forty years, he wrote at least three full length novels every year. Over 80 of these featured Perry Mason, and the last one, *The Case of the Postponed Murder*, was published posthumously in 1973. He had also penned over 50 other detective fiction books, about 30 of which starred Bertha Cool, a hefty woman private investigator, which Gardner wrote under the pen-name of A. A. Fair.

In 1946 Perry Mason began appearing in his own comic book. According to comic book historian, Mike Benton, this first publication portrayed Mason as a tough lawyer, willing to use his fists to settle a problem, not unlike the attorney Gardner wrote of in the early novels. Unlike the calm, urbane, and professional lawyer of the later novels or the popular television series, the first comic book Mason thrived on physical confrontation. However, in 1964 another comic book series began, spurred by the enormous success of the television program. This time Mason kept his temper and his fists in check. His attitude resembled the low-key interpretation of television actor Raymond Burr, whose photograph frequently appeared on the comic book covers.

The radio series of Perry Mason began in October 1943 on CBS and it remained on that network for the next twelve years. It was, like *Kitty Keene, Incorporated* and *Front Page Farrell,* a hybrid radio program combining soap opera and detective adventure. Over 300 audio episodes have survived to the present day, although not all of them contain the commercials of Proctor and Gamble, the longtime sponsor, extolling the benefits of Tide detergent and Camay soap. As in the novels, Paul Drake, who ran a detective agency, and Lt. Tragg, of the local police, were usually in the radio stories, helping or hindering Mason.

While the network did nothing to discourage listeners from assuming that Erle Stanley Gardner actually wrote every script, virtually the only writing he did in connection with the radio series was to sign his name on the royalty checks. The majority of the approximately 3,000 programs aired were written by Ruth Borden, Dan Shuffman, Irving Vendig, and/or Eugene Wang.

Since this radio series had a long tenure, several different radio performers played the co-leads of Perry and Della. The combination that had the longest time at the microphone, almost eight years, was the pairing of Joan Alexander and John Larkin. Other combinations that made up this duo over the years included the voices of Santos Ortega, Donald Briggs, Bartlett Robinson, Gertrude Warner, and Jan Miner.

All of the above were both talented and successful; most of them were also leads on other radio series. Joan Alexander was the female lead on *The*

Adventures of Superman, It's Murder and *The Man from G-2*. Jan Miner, in addition to Della Street, was also the voice of Annie Williams on *Casey, Crime Photographer*. Gertrude Warner worked on several soap operas (including *Against the Storm* and *David Harum*), as well as anthologies (*Brownstone Theater* and *Matinee Theater*). Warner, in addition to her role as Mason's secretary, also starred as the feminine lead in three other crime dramas, *The Adventures of Ellery Queen, The Shadow,* and *City Desk*.

Both Donald Briggs and Santos Ortega were accustomed to being the principal in a radio series devoted to crime-solving. Briggs played the leading role in *The Adventures of Frank Merriwell, The FBI in Peace and War,* and *The Sheriff*. Ortega was one of the busiest actors on network radio and specialized in portraying detectives, some with a distinctive accent. He was the voice of Charlie Chan (Asian), Hannibal Cobb (rustic), Bulldog Drummond (British) and Nero Wolfe (cultured.) While Bartlett Robinson could not claim any major roles, except Perry Mason, he was a skilled radio actor and had done well on several soap operas, including *Portia Faces Life, Valiant Lady,* and *The Romance of Helen Trent*. Fans also remember him as "Rupert Barlow," Larry Noble's rival on *Backstage Wife*.

There was a marked contrast in the relationship of Perry Mason and Della Street in the original novels of the 1930s, when compared with their radio counterparts. In Gardner's books, it was clear that Perry and Della were romantically involved. They snuggled in his automobile while on a surveillance, occasionally shared a drink from the flask in his glove compartment, and kissed goodnight fervently at the end of each tough day. Gardner makes it apparent that she was sexy ("Della, with a flash of legs and flounce of skirts, slid across the car seat"), risk-taking (every time Perry tells her not to follow him into danger, she does) and mischievous (after a long kiss, she departed, saying "Remember to wipe the lipstick off, Chief.")

The Della of Gardner's novels is the intellectual equal of Mason, and sometimes his superior. Describing a female client whom Mason had not seen, Della admonished him, "I'd place her at twenty-six. You'd guess her age as twenty-one." During one case, she and Mason were surreptitiously searching an apartment for a diary. She not only found it first, she quickly concealed it from the police who had just broken into the apartment. Della rolled up the diary, shoved it into a loaf of bread, which she then dumped in the trash, where she and Mason retrieved it, after the police ended their search for the same item.

There are even instances in the novels that would lead the reader to believe Della and Perry are so attuned to each other, they can almost read each other's minds. In one particular mystery, Della telephoned Perry, while he was in the company of a police officer. Wanting to exchange vital information with Della, without the police learning of it, Perry made up a code on the spot, and Della understood the entire message.

In Gardner's novels of the 1950s and 1960s, two significant changes occurred. Perry spent a great deal more time in the courtroom so prosecuting attorney, Hamilton Burger, got more pages than Lt. Tragg. Secondly, the romance of Perry and Della cooled down to the vanishing point, with no kissing, no snuggling, and no sharing the flask in Perry's car. In these later novels, Della was promoted to "confidential secretary," although her duties were identical, and Gardner described her in strictly professional terms, "Della, trim and efficient, with leather briefcase in her gloved hand..." In the novel, *Case of the Screaming Woman* , Perry was teased into complimenting Della and so he stammered out: "It's nice to have you around...It's nice to have your loyalty, your dependability, and...you're easy on the eyes."

Perry not only brought Della on all surveillances and interviews in *The Case of the Glamorous Ghost*, he also used her for bait to snare a jewel thief. For this undercover work, he put her up in a posh hotel and instructed her to make herself available at the pool and on the dance floor. Apparently Perry thought she was still physically attractive to other men, even though he had lost interest.

Radio's Della Street was certainly as competent, dedicated, and courageous as the one in the novels. But instead of being sexy, she was a modest and professional business woman, much as the Della of the later novels and the television series. Any romance with the boss does not figure in the radio adventures, except in the vaguest of terms. And if either Perry or Della reached for the flask in the glove compartment, it was to prevent frostbite, not kindle affection. In the novels, Della spent virtually no time in the office doing secretarial work as she was always with Perry on a surveillance, witness interrogation, or gathering clues. The radio scripts gave Della fewer assignments outside the office.

One reason for the difference in Della's behavior is the medium: in 1940s radio, a secretary could smooch with her boss only in a comedy. In a "realistic" crime drama, most scriptwriters stayed away from sexy encounters between responsible adults. (Of course, the married couples, i.e. Nick and Nora Charles, could get away with more.) Another reason would be that not only was *Perry Mason* airing in the traditional soap opera afternoon period, but some years, depending on who was churning out the scripts, it could be mistaken for any other soap opera on the air. So Della's character and behavior had to comply with the somewhat artificial standards that guided the women's daily dramas. Occasionally the *Perry Mason* series veered so deeply into its soap opera venue that Perry and Della almost disappeared. An example of this, taken from an episode in September 1949, appears below.

This episode began with the customary Tide commercial, approximately a minute and a half, and then the scene opened with Mason and Lt. Tragg at the door of the suspected hideout of murderer, Bill Barker. As soon as they demanded entrance, the announcer jumped in...

ANNOUNCER: Mason and Tragg are certain Barker's inside, and he's well-armed—and desperate! Well, we'll come back to Tragg and Mason in just a few moments, but first...in the living room of the Tragg home, Tragg's wife stands listening to a radio that's tuned to the police calls...stands quietly as she hears...

FILTER MIKE: (*Bulletin about Tragg closing in on Barker*)

MRS. TRAGG: I was just listening to the police bulletins. Now what about Nora?

MARTHA: Just that she's finished her egg nog and settled down to take her nap. The rate she's going, she'll be fine, out of bed in a few days.

MRS. TRAGG: Thank you, Martha

MARTHA: But I'm afraid I'm not able to say the same about you.

MRS. TRAGG: But why?

MARTHA: Sit down here, and let me get you an egg nog.

MRS. TRAGG: Oh no...no thanks.

MARTHA: You can worry just as well on a full stomach as an empty one. I learned that in the war, when I'd be without news of Dick for months at a time. Every time the doorbell would ring, I'd worry...Well, I couldn't go without eating forever and I couldn't stop worrying, so I learned to worry and eat.

Now…can I get you an eggnog?

MRS. TRAGG: No…Your husband was killed, wasn't he?

MARTHA: In 1944. We were married in 1942. We really never got a chance to know each other. He went overseas right after the honeymoon…he never came back.

MRS. TRAGG: Oh, I'm sorry.

MARTHA: It's OK; I don't mind talking about it…

The conversation continued between the two women for the next ten minutes as they discussed whether Martha should ever remarry, how close as neighbors they've become, the absence of gossip in their block, if Lt. Tragg is in danger, the problems of being the wife of a policeman, and what should they make for dinner? In the closing seconds of the episode, the announcer took the radio audience back to Barker's hideout, where Tragg and Mason broke in and find the killer has escaped out the back, followed immediately by the Proctor and Gamble closing commercial.

During other years, under the hands of different scriptwriters, *Perry Mason* was divested of most of the soap opera trimmings and became primarily a crime adventure show. One mid-40s story line involved a ruthless killer, Gus Jansen. In an interesting example of casting against type, Jansen was played by Bret Morrison, who was then the voice of Lamont Cranston and The Shadow. The killer was on the trail of a fleeing blonde, and over several weeks' programming, Perry, Della, and Paul Drake tried to locate the intended victim before Jansen could find and kill her.

At one point, Jansen located the rooming house which the young lady had just vacated. Perry and his assistants were one step behind the murderer, checking drugstore records across town. After Jansen extracted the needed information from the frightened landlady, he brutally shoved her down the cellar steps to her death. He managed to flee the crime scene, just before Mason and his two associates arrived at the rooming house. Perry instructed Della to stay in the car so he and Drake could make sure no danger lurked in the building. True to form, Della ignored his orders and entered the house just as Mason and Drake discovered the corpse.

Another perspective from which to examine the *Perry Mason* series, including the relationship of Perry and Della, is in the illustrated pages of *Radio Mirror* in the 1940s. Each month that periodical would highlight a different radio series, usually an adventure show, a soap opera, or a mystery. The program chosen would be featured in a two-page spread, consisting of about eight photographs, plus a small amount of text, based on a current or recent radio script, which was then modified to fit the article. Half-hour shows, which resolved a mystery within one program, were easier to adapt than serials like *Perry Mason*, whose 15-minute show could involve 30-40 episodes before the conclusion of the story line.

Once the text was approved, *Radio Mirror* made arrangements for the radio performers to pose for a series of indoor and outdoor photographs to illustrate the story. In the June 1947 issue, Perry Mason's *Case of the Bartered Bride* was featured. Donald Briggs and Jan Miner (described as " his bright-haired secretary who is fearlessly, wholeheartedly involved in all of his cases") were joined by their fellow radio actors in the black-and-white illustrations. This case began with Della, ignoring Perry's orders, went to the Mirado Cafe to look for the coat of accused, but innocent, May Blade. Della was later kidnapped by the murderer, who then held her hostage. The story ended as Mason overcame the killer and rescued his secretary. The last caption read: "Della is an extremely important person to Perry, though he may not be able to say why. Maybe, one day, Della will tell him."

About a year later, in its July 1948 issue, *Radio Mirror* again featured *Perry Mason* in its monthly illustrated story. This time the radio leads photographed for the spread were John Larkin and Joan Alexander. In this mystery, *The Case of the Sinister Sister*, Della is instrumental in helping Perry solve the crime. Then, in the closing photograph, Alexander peers wistfully into the camera, above the caption: "Della hopes Perry will one day see her, not just as an invaluable co-worker, but a woman to love."

MRS. MULLET

Though bereft of a first name, Mrs. Mullet served as housekeeper, confidant, and mother hen to amateur detective, Leonidas Witherall. While she is prominent in their radio adventures, she is nearly invisible in his mystery novels. The presumption would be that she was created as a feminine companion to Witherall, when he transitioned from the printed page to network radio, much the same as Kay Fields was added to Dan Dunn's company for his radio adventures.

Mr. Witherall, a fictional crime-solving gentleman, was the product of the inventive mind and prolific pen of Phoebe Atwood Taylor (1909-1976), a New England mystery writer. Beginning in 1931, she wrote twenty-four novels featuring a sleuth named Asey Mayo. This Massachusetts character, a former sailor and auto racer, successfully solved two dozen baffling cases, starting with *The Cape Cod Mystery* (1931) and continuing through to *The Diplomatic Corpse* (1951).

Leonidas Witherall is actually considered by mystery fans to be "her other detective." Using the nom de plume of Alice Tilton, (misspelled in some radio reference books as "Alice Tilden") she wrote about a half-dozen mystery novels around Witherall, beginning in the late 1930s and ending with *The Iron Clew* (1947.) Both of her fictional sleuths were oddly-named, intelligent, humorous, eccentric, and very popular with the public. Most of her books were reprinted as paperbacks in the 1980s, after her death, and copies of her original hardbacks command high prices at Internet sources of used books, as do those of the Abbott mysteries of Frances Crane.

In Taylor's novels, Leonidas Witherall was the headmaster of Meredith Academy (although it was commandeered by the U.S. Navy in World War II), served on prominent community boards, and lived alone in a large house in a Boston suburb. He was an expert on William Shakespeare, quoted him often, and resembled him physically to such an extent that friends, and strangers alike, frequently addressed him as "Shakespeare" or "Will." And while the Bard of Avon was probably never called "Bill," Witherall was. He paid for his lavish residence with the money received from writing a radio crime show starring a Lt. Haseltine. This may have been an inside joke by Taylor, who certainly knew that most radio writers of that era earned far less than successful mystery novelists.

The Adventures of Leonidas Witherall debuted on Mutual in June 1944, emanating from the studios of WOR in New York City. It was a half-hour program on Sunday evenings and ran until May 1945, a sustaining program that never found a sponsor. Copies of seven different episodes from this series have survived to the present day.

While creator Taylor had little to do with the radio version, she must have been pleased that virtually all of the characteristics and avocations of her hero were copied from her books into the network show. Witherall's producing the Lt. Haseltine stories was shifted to a minor key on the radio and only mentioned infrequently, after the standard opening of each episode. In Taylor's novels, the subject came up frequently; one character recited the opening of that mythical program: "Did you ever listen to the radio version that begins with thunder and bugles and a man yelping 'HASEL-tine to the RES-cue!'?" In another scene from her books, a character reminds Witherall that Haseltine once disguised himself as a tree, to cross an open field without being detected.

Since radio's Witherall was in constant company of a female, he had to discontinue the occasional sarcasm he directed at women. During one of the mid-40s mystery books, Witherall and another man and woman were locked in an anteroom by an evildoer. Witherall could not find any suitable tool to begin their escape until the woman handed him a sturdy screwdriver, retrieved from her jacket pocket. Instead of thanking her, the sleuth retorted with some condescension. "The millennium has arrived! At last, I have seen a crisis where a woman did not bring forth a nail file and expect one to move mountains with it."

At WOR, there was no doubt who was the star of this scholarly crime-solver series; the lead, Walter Hampden, appeared above the title. Here's the opening of the show, heard by both the live audience in Manhattan and the listenership throughout the nation:

ANNOUNCER: WOR, Mutual, presents the distinguished American actor, Walter Hampden, in…The Adventures of Leonidas Witherall!

FIRST MAN: Leonidas Witherall is always getting mixed up in murder!

SECOND MAN: Well, you wouldn't think so; he looks just like Shakespeare.

WOMAN: It's his beard. And he's head of an important school for boys in New England

FIRST MAN: He also writes thriller stories on the side too, Lt. Haseltine stories.

SOUND: [Organ sting]

Agnes Moorehead was originally chosen to co-star with Hampden as his housekeeper, Mrs. Mullet. Vintage radio historian, Jim Cox, has advised that the two performers made an audition recording of this radio program as early as September 1943. It is unclear how many times Moorehead played the role, but Ethel Remey took over early in the run and portrayed Witherall's housekeeper until the series ended. Remey had few major roles, other than this series, although she was in the cast of Young Widder Brown. It's likely she was chosen for this part because her voice closely resembled Moorehead's. And for Mullet, Remey produced a high, wavering voice typical of someone's elderly aunt.

Hampden had even less radio experience than she did. Primarily a stage actor, he made few forays to the radio microphone, other than briefly hosting Mutual's anthology, *Great Scenes from Great Plays* in 1948. He played Witherall as an astute, eccentric, and jovial gentleman; he made his character seem to be equally comfortable riding a bicycle or driving a horse-drawn buggy.

There is some confusion about the spelling of his housekeeper's name. Most radio reference books cite her as "Mrs. Mollet," but in the audio copies of this series, all the characters clearly address her as "Mrs. Mullet." The fact that a woman should bear the name of a spiny-rayed fish was not unusual in the historical tradition of Leonidas Witherall. His creator loved coming up with peculiar names to populate her mystery books. For example, in Witherall's 1943 novel, *File For Record*, she places the following: Mr. Haymaker, Mrs. Pink Lately, Mr. Gidding, Elsbeth Agry, and Major Breeding. The radio scriptwriter, Howard Merrill, cheerfully followed her custom. In the September 24, 1944 episode, entitled "Murder at the Country Fair," Witherall and Mullet meet: Mrs. Axlebent, Mr. Whacker, Miss Maplegrove, and Mr. Frivet.

Leonidas Witherall not only resembled Shakespeare and quoted him liberally in the mystery novels, he was careful to use Elizabethan pronunciation, i.e. breaking "topped" into two syllables. He also employed a certain theatricality in solving the mysteries thrust at him. To gather evidence in one case, he had a young woman friend fake a fainting spell and then he brought in a confederate, attired in a chef's white coat, and passed him off as a doctor to diagnose the young woman as suffering from a sulfur overdose. Scriptwriter Merrill picked up this technique and used variations of it in the radio show. In an episode, misleadingly titled "Peaceful Vacation," the plot involved Witherall getting off a pier into a lake, and wearing water-wings to conceal the fact that he was an excellent swimmer. This ruse lured an unknown murderer to throw a knife at the water-wings, assuming Witherall would drown.

Mrs. Mullet thrived upon the excitement of sirens on passing emergency vehicles, loved all neighborhood gossip, and was especially delighted to assist the headmaster in solving any murder or puzzling crime. He gently, and smilingly, accepted her aid, regardless of its potential import.

MULLET: So, we're going to look into a murder, eh? I'll get to
          the bottom of this. Where's my hat? Where's my coat?

WITHERALL: Yes, and don't forget your bloodhound.
(*Chuckle*)

As might be supposed, the police are not as accepting of Mullet's tagging along with her employer. One police officer chastised her: "It's time for spring cleaning; why don't you go home and rearrange the dirt?" She retorts by calling him "Squeaky Shoes." Another time, a constable, out of her earshot, tells Witherall, "Mrs. Mullet is the only over-aged destroyer that wasn't sent to Britain."

Most of the radio scripts provide Leonidas Witherall and Mrs. Mullet with an astonishing variety of crimes, in terms of unusual weapons of death. In three different episodes, the victim was dispatched by hot, molten lead from a Linotype machine, stabbed in a dark movie theater, and placed bound and gagged between a target canvas and the metal back wall of a shooting gallery. The latter method was a second choice for the killer when the original plan failed; he had used a poison-filled hypodermic needle to penetrate the paraffin top on a jar of preserves.

Through it all, Witherall and Mullet persevered, and enjoyed each other's company, from the first discovery of each crime to the solution of the mystery. While there were sporadic mentions of a "Mr. Mullet," her spouse never appeared in their radio adventures. That may have accounted for the confusion of the carnival barker in one mystery, who seeing Witherall and Mullet near his booth, asked him: "Wanna win a Kewpie doll for your girlfriend?"

Although prone to bursts of excitement, Mrs. Mullet was a thoroughly dedicated, considerate, and helpful associate. In one thrilling episode, she was alone and found a bloody corpse in a mountain resort. She calmly located Witherall with a group nearby, remained attentive at his side until a lengthy conversation ended and only then, quietly reported the murder to him.

Mrs. Mullet may well have been the oldest character, among all the lady crime-fighters of the Golden Age of Radio, but with her grit, spunk, and energy, she was the equal of many of her juniors.

BARBARA SUTTON

In July 1943 *The Black Hood* arrived on the Mutual Network, coming straight out of the pages of the comic books and pulp fiction magazines where he had been featured since 1940. His feminine companion, Barbara Sutton, a newspaper reporter, ably assisted him in combating crime and thwarting assorted thugs.

There were many dashing crime-fighters in the comics with "black" in their name: *The Black Owl, The Black Terror, Blackhawk, The Black Condor,* and even a woman, *The Black Cat.* But by far the most ruthless was *The Black Hood,* a character created in October 1940 by Harry Shorten, a writer and editor for the fledgling organization, MLJ Comics. Although the origin story was not unique (i.e. good cop, framed by evil forces, turned into a secret vigilante), *The Black Hood* caught on with its readers, and within a year had branched out into his own pulp fiction publication, *Black Hood Detective Magazine.*

The character began as a lowly police patrolman, Kip Burland who, while trying to stop a burglary at a mansion, was knocked out by "The Skull," a criminal disguised in green skeleton attire. The Skull then planted some stolen jewelry on the unconscious Burland, and escaped. By the time Burland recovered, he was under arrest. After making bail, he tried to apprehend The Skull, but his nemesis blasted him full of holes and discarded the near-dead body in a forest river. Fortunately, he was found there by an old hermit who not only restored him to health, but also helped him achieve super strength and learn all scientific knowledge for his crusade against crime.

In order to conceal his identity from both criminals and the police, Burland donned a "black" hood as part of his crime-fighting attire. However, as drawn by successive artists Al Casmy, Bob Montana (who in 1941 would begin *Archie Andrews*), Charles Biro, and Irv Novick, his costume consisted of a canary yellow leotard, over which he wore a dark blue hood, trunks, gloves, and musketeer-style boots, but not a cape. Apparently, dark blue was better for showing details in the comic pages than pure black. Later, as The Black Hood, Burland successfully captured The Skull, but it was a short-lived victory as Burland's supervisor, Sergeant McGinty, then threw The Black Hood behind bars on suspicion charges.

Since this was a comic book, no one thought to take the mask off The Black Hood in jail and discover police patrolman Kip Burland underneath. But he was saved from confinement through the efforts of Barbara Sutton, a reporter for the fictional *Northville Courier,* whose local news articles resulted in the freedom for The Black Hood. Burland gratefully confided in Sutton, regarding his dual identity, and the two forged a semi-romantic alliance to continue the fight against all criminals.

For the next five years, Barbara Sutton marched across the comic books pages with Kip Burland, with or without his black hood. She was portrayed as a trim, athletic brunette with very short hair, smart clothes, and usually wearing a pillbox hat. Again, since this was a comic book, no one ever figured out that Burland was The Black Hood, even though Sutton was always with either one.

While this comic book heroine was occasionally put in a perilous position by any of the sundry and colorful villains which The Black Hood pursued, including Panther Man, The Mist, and The Crow, she was more often holding her own against these thugs. A typical scene appears in Black Hood comic # 14 where The Black Hood crushed one hoodlum in a door frame but a second punk gets the drop on our hero. Sutton, who was behind the gunman, slugged him into oblivion with her "recording machine," yelling as she does, "Never turn your back on a lady! It's not polite and sometimes, it's plain dumb!"

Harry Shorten was joined by other writers, including Bill Woolfolk, in keeping The Black Hood knee-deep in dangerous situations, several of which involved strong violence, even bondage and torture. This comic book crime-fighter seemed to rely more on bone-crushing strength than scientific knowledge in capturing evildoers. With Sutton at his side, he lasted in his own comic book until the summer of 1946, with a final appearance in *Pep* magazine when the era of the super heroes was winding down. But he was certainly at the height of his popularity when he arrived on the airwaves.

The Black Hood first reached its juvenile listening audience on July 7, 1943 on Mutual, in the middle of World War II. It would last until the following January, but never attracted a sponsor so it was a sustaining program for the entire run. It was a serial, with each program running 15 minutes, five times a week, usually at 5:15 PM Eastern. The initial broadcast opened, as all the rest would, with an Oriental gong sounding, followed by actor Scott Douglas, as The Black Hood, growling into the microphone the oath that had first been developed in the comic book:

I, The Black Hood, do solemnly swear that…neither threats, nor bribes, nor bullets, nor death itself…shall keep me from fulfilling my vow: to erase crime from the face of the earth!

Scott Douglas, playing both Kip Burland and The Black Hood, was joined at the microphone by Marjorie Cramer, who was the voice of Barbara Sutton, usually called "Babs." These leading roles may have been the high point of their respective careers since neither one is mentioned in any other program in all standard radio references. But both were convincing and suited their parts well, which were slightly modified from the comic books and pulp magazines. The unusual villains remained, but the violence was toned down, with the incidents of bondage and torture virtually eliminated.

The theme music was a strange choice: *L'Apprenti Socier"* (or *The Sorcerer's Apprentice*) by French composer Paul Dukas. Although this composition was first performed in 1897, it became well known through the popularity of Walt Disney's 1940 film, *Fantasia,* where it served for the background music for Mickey Mouse's segment with the water buckets being carried by walking broomsticks. Whether or not many members of the juvenile radio audience recognized the tune and connected it to Mickey is unknown. But the music was probably chosen by the radio producers because it was unusual, energetic, and best of all, royalty-free.

In addition to Burland and Sutton, one more character from the comic book was included in the radio series, Sergeant McGinty, Burland's not-too-bright police supervisor. The three interacted in virtually every episode, with McGinty usually requiring assistance, and sometimes, rescue. Every bridge between scenes in each program was announced by the Oriental gong sound, which must have become irritating to the cast after four or five such annoying crashes.

Only one episode has survived of this adventure series, but it is probably characteristic of the rest of the run, since it showcases the strength of The Black Hood, the courage of Barbara Sutton, and the ineptitude of McGinty, complete with exotic villains. Although it was the debut program, the uncredited announcer began by explaining what has happened "before": Kip and Barbara had visited a 90-year-old voodoo doctor, known as the Miracle Man, who was attended by a weird housekeeper, Wamba. The Miracle Man gave Barbara a mysterious, emerald, serpentine ring which she wore home. Then the drama began with an armed intruder confronting Barbara in her residence while she was on the telephone with Burland. The masked robber demanded money, valuables, and then the emerald ring. Although she had his gun in her face, Sutton refused and struggled valiantly with the gunman. Kip (who was apparently in a telephone booth just outside her house?) changed into The Black Hood, and two seconds later, crashed in, just in time to see the robber escape.

The scriptwriting was a little ragged, and The Black Hood, after examining the crime scene and finding cash the robber dropped, announced: "Any good, self-respecting robber always puts stolen money in his pocket, so in a quick getaway, he gets away with something." Sutton agreed. Next day she reported the robbery to Sergeant McGinty, in the presence of Burland, who's also on duty. McGinty suspected the Miracle Man of the crime and made plans to visit him alone that night.

In the next scene, we learned that the Miracle Man, through the power of voodoo, is ridding the world of undesirables. The previous night, he killed a man who was a wife-beater, and the Miracle Man was planning to use his voodoo to murder another troublemaker this evening. Meanwhile, Kip and Barbara were parked in his car in the moonlight, not smooching, but examining the emerald ring. She discovered the top opens up, and inside, Kip observed some powder. Since this was 1943, their first thought was not cocaine, but poison, so Kip promised to have it analyzed in the police laboratory the next day. He then invited her to join The Black Hood on a "secret midnight mission" and she eagerly agreed.

A few miles away, Sergeant McGinty had bullied his way into the voodoo residence of the Miracle Man, despite the protests of Wamba. She pretended to show him some African crockery, and while he was distracted, Wamba smashed it over his head, rendering him unconscious. So at this juncture, following yet another crash of the Oriental gong, the announcer breathlessly asked the kiddie audience:

> "Now, what will Wamba do with the unconscious McGinty? And what about The Black Hood—has he been able to discover anything more about the emerald ring and its strange powder? If Wamba intends to harm the good sergeant, can the Black Hood and Barbara arrive in time to save him? Be sure to listen tomorrow for more thrilling adventures of The Black Hood!"

The program then ended with another bouncy section from *The Sorcerer's Apprentice.*

JUNE SHERMAN

The broadcast history of *Crime on the Waterfront* is a trifle murky; it's possible that NBC never actually aired the two episodes, copies of which are being traded among collectors today. But there is also a rebuttable presumption that both shows were broadcast, respectively, on February 24 and March 1, 1949. In either case, June Sherman appeared in only one of these episodes, and it is clearly superior to the one without her.

Fans of *Bulldog Drummond* must have noticed the similarity in the openings of *Crime on the Waterfront* and their favorite series. Both started with a deep foghorn, followed by the sound of the hero's footsteps on a wet, hard surface. However, at that point, instead of shots being fired and a blast of a police whistle, *Crime on the Waterfront* switched to the sounds of coins dropping in a pay telephone and our hero's voice: "Waterfront...Kagel calling." This was the announcer's cue:

> "The National Broadcasting Company presents: Lew Kagel,
> ace New York detective...fighter of crime on the waterfront!"

Radio listeners, hearing the first episodes in 1949, had reason to be somewhat confused since George Stone was the announcer, not Myron "Mike" Wallace, who was Kagel. In the Golden Age of Radio, men coming to the microphone were classified as either actors or announcers, and some fellows, despite years of trying, never succeeded in moving from one category to another. Wallace, now the dour-faced, investigative journalist that today's television audience watches on CBS's *60 Minutes,* spent most of his radio career as an announcer.

At various times, Wallace was the announcer on *The Green Hornet, Curtain Time, A Life In Your Hands, The Spike Jones Show, You Bet Your Life,* and *Sky King.* On the latter series, he had to exuberantly encourage the juvenile listeners, in all the commercials, to eat Peter Pan Peanut Butter. Mutual finally gave Mike a chance to act, casting him in the lead in *The Crime Files of Flamond* in 1946, produced at WGN in Chicago, possibly his first time at the microphone when he was not announcing.

In *Crime on the Waterfront,* he was Lt. Lew Kagel, a tough but tender police detective, assigned to Manhattan's harbor area. Although some standard radio reference books spell his first name as "Lou," it is certain that "Lew" is correct because his girlfriend, June Sherman, a newspaper journalist, occasionally teased him by using his given name, Llewellyn. In turn, he sometimes refers to her as "Missy" or "Junie."

June was portrayed by Muriel Bremner, whose experience on network radio up to this time was fairly well confined to the soap operas at NBC and CBS. She was in the supporting casts of *The Guiding Light* and *Road of Life.* She and Wallace both had leading roles in *The Crime Files of Flamond,* but Mike left that cast in 1948 and Muriel did not join this series until it was resurrected in the mid-1950s.

Kagel described June to his radio audience as: "A frantic young lady with a press card in one hand and a pad and pencil in the other." Later he confided, "Occasionally, we have talked about marriage" so this was clearly a serious relationship. And in his criminal investigations, they were joined at the hip, just like radio's married couples in crime, Pam and Jerry North, Nick and Nora Charles, etc.

Lew and June were together all the time, whether it was business or pleasure. They interrogated suspects together. They went to police headquarters and jointly read law enforcement agency teletypes and ballistic reports from the crime laboratory. Later, they took the evening off, attended a hockey game, followed by burgers at a downtown hamburger joint. Later that night Lew got word of new developments in their case, so they split up, each taking a separate part of the inquiry to resolve.

The scripts, by J.T. Kelly, had crisp, logical plots that advanced the action well. Other production values, including sound effects, were handled by very competent personnel. Wallace and Bremner were backed up by a strong supporting cast. While the director is not credited by announcer George Stone on either existing audio copy, it is evident that whoever it was knew their craft.

Throughout their adventure, June was resourceful, disciplined, and eager to solve the case, not for a newspaper scoop, but rather taking pleasure in the triumph of justice. In short, she was everything Lew, or any other law enforcement officer, would want in a police partner...or a prospective spouse. By the end of the episode, most of the radio audience had forgotten that June was a reporter, and probably, so had Lew.

Well, Just For Laughs

If it is true that laughter is the best medicine, then some of radio's feminine sleuths dispensed healing remedies with their zany deductions. Creating a humorous network series based upon a lady crime-solver was not an easy task and the four programs discussed in this chapter demonstrate varied degrees of success in the endeavor. At least two of them were usually funny, one missed the mark entirely, and an accurate assessment of the fourth will have to wait for the discovery of a sample of its scripts or audio copies.

SARA BERNER

Sara Berner, a pint-sized powerhouse performer, was given the starring role in the NBC summer series, *Sara's Private Caper,* in which she used her own name. The only other lady detective to do so was Irene Delroy in the thirties. While Sara's series masqueraded as a detective program, it was a pure comedy show.

Berner was born in September 1912 in Albany, New York, but her family moved to Oklahoma while she was still in grade school. On weekends, her parents gave her money to take her younger brother to the movies in Tulsa. She fell into a custom of dropping him off at a theater that played cowboy films, and then she went to a vaudeville theater that had live acts and a silent movie. Little Sara was bitten by the acting bug early and she loved to do impressions of the acts she had just seen, for the theater patrons as they exited.

Later, she was active in high school drama as well as a Tulsa community theater. About the time she graduated from high school, the family moved again, and Sara found herself in Philadelphia. She promptly got a full time job as a retail clerk at Wanamakers Department Store, while volunteering for any radio assignment at Columbia University's station, WCAU. At Wanamakers,

167

Sara Berner of *Sara's Private Caper* in her Jack Benny years (*The Stumpf-Ohmart Collection*)

Berner entertained co-employees by doing impressions of unpleasant customers, however her manager was not amused and she was fired.

By that time WCAU was giving her significant roles, including a 15-minute program of her own, written by Arthur Q. Bryan. But she needed to earn a salary, so she went to New York City, found employment at a Broadway ladies hat shop, and started making the rounds of radio stations and advertising agencies, trying to get work as a radio performer.

She managed to talk her way into getting a slot as a contestant on the Major Bowes Amateur Hour and after she won, she was selected to tour with the Bowes theater troupe. The experience she gained there insured her of a successful career in radio when she eventually returned to Manhattan. Berner was one of several women who played "Miss Duffy" on *Duffy's Tavern* and she played "Bubbles" in its summer replacement, *Nitwit Court*. Her skills in mimicry and impressions led to regular roles on *Eddie Cantor Show, Baby Snooks,* and *Burns and Allen.*

Berner later concentrated her radio work on the West coast, with her parade of wacky characters with unusual accents. Respected radio historian, John Dunning, credits her as "Mable Flapsaddle and Gladys Zybysco" with Jack Benny, "Ingrid Mataratza" with Jimmy Durante, "Mrs. Horowitz" on *Life With Luigi,* and "Mrs. Jacoby" with Dennis Day. She was also playing Mexican senoritas on Gene Autry's show.

Her Los Angeles base gave her access to Hollywood's animated film production and she frequently voiced characters in cartoons, including *Little Jasper.* She was so successful that five of the animated movies in which she did principal vocalization won Academy Awards, including *Red Hot Riding Hood* and *Mother Goose Goes Hollywood.* Probably the pinnacle of her cartoon work occurred when she supplied the voice of the animated mouse who danced with Gene Kelly in *Anchors Aweigh.* Her exuberant cry, "Look at me; I'm dancing!" became a popular expression in the American popular culture.

So by the time this 5-foot, 3-inch performer, who barely weighed 115 pounds, stepped up to the microphone in June 1950 to begin her own detective parody program, she was already a successful and well-known personality. General Mills sponsored her series for the three months it was on the air. Since the show was broadcast late on Thursday evenings, it had a largely adult audience, so the writers of the Wheaties commercials struggled in vain to convince grown-ups to consume what was generally considered to be a kids' cereal.

Berner played a police secretary, who was under strict orders by her supervisor, Lt. Phillips, to limit herself to typing and filing reports, followed by sharpening pencils. Each episode began with Sara listening to a radio thriller. In the two surviving shows we have now of Sara's series, Howard Duff in *The Adventures of Sam Spade, Detective* and Frank Lovejoy in *Nightbeat,* were heard in brief transcriptions. The remainder of each plot outline of Sara's program was akin to a scrawny Christmas tree, upon which ornaments of Berner's vocal characterizations were hung. Most of the programs were indistinguishable from one another, except for the variety of her different dialects and accents.

In each program Sara found a police report of a new crime, usually the theft of property. Ordered not to leave the office or investigate the case, she immediately did just that. She then joined her boyfriend, Melvin, who was a clerk at the United Food Market, where they usually knocked over a pyramid of canned goods, to the consternation of his boss, Mr. Sacks. Through a series of disguises and clever vocalizations, Sara solved each caper, just in time for the announcer, Frank Martin, to read the Wheaties closing commercial.

While the above summary of the standard show may suggest this production was done cheaply, it certainly was not. The program was done in front of a studio audience, and joining the cast and sound effects personnel on stage, was a full orchestra, conducted by Robert Armbruster, who composed all the music. In the early 1950s, most network shows had downgraded live music down to one organist, or even did away with it in favor of transcribed music, so General Mills was shelling out big bucks for this orchestra. Another way of measuring the production costs was the size of the cast. During this era, most of the producers for network comedy and drama shows, when faced with a script that called for eight supporting characters, would hire four and make each do two different roles. Not so with *Sara's Private Caper*. In the episode of the missing book, in addition to the four regulars in the cast, eight supporting performers were utilized: Eric Snowden, Gerald Mohr, Tony Barrett, Peter Leeds, Jerry Hauser, Tom Holland, Donald Morrison, and Jack Petruzzi.

Although Joe Parker, the producer and director, contributed to the writing of most scripts, he had plenty of help in this department. Putting Sara's jokes on paper was also done by Ben Starr, Larry Klein, Morton Fine, and David Friedkin. If their one-liners weren't that hilarious, Sara could still get a laugh with her accent or facial expression. And some of the jokes weren't that great. When she gave a client a wrong answer from her crystal ball, Sara added quickly that she had the wrong channel. After another mistaken prediction in the ball, she claimed it was cloudy from the Hollywood smog. Then, when she tried to fool a suspicious character:

SARA: Oh, then the book was *stolen!*

OWENS: Who said it was stolen?

SARA: I read in the paper that a lot of books were stolen. The headline was "BOOKIES SNATCHED BY MOB."

OWENS: Bookies?

SARA: Yeah, must have been itsy-bitsy books, huh?

As in other funny radio shows done on stage with a studio audience, there was obviously a lot of strictly visual comedy that the radio audience missed. The uncredited sound effects personnel received a great deal of laughter, unrelated to the actual sounds produced. The script always called for Sara to flirt with one or more characters on the program, which resulted in hilarity during what were pauses in the script. Either Sara's mugging, or the disparity in height of her and any man at the microphone, was enough to nudge her studio audience into gales of mirth.

Since this was a parody of a detective show, our heroine did not have to be intelligent, courageous or logical, and she was not. But as dizzy as Sara was in each episode, she appeared almost brilliant, compared to her boyfriend, Melvin. A lowly stock boy, who aspired to be an assistant manager at his store, he was played by Bob Sweeney. He was only about 30 when he got this role, but had had some past success in radio as the nervous half of a comedy team, airing as *Sweeney and March* ; his partner was Hal March. Their variety sketch program was broadcast sporadically by CBS at various periods from 1946 to 1951.

Sweeney had supporting roles in other West Coast shows, including *Halls of Ivy* and *Cousin Willie*. One of his claims to fame was portraying the co-lead in NBC-TV's *Fibber McGee and Molly*, with Cathy Lewis, whom he had worked with on her radio anthology series, *On Stage*. Radio's *Fibber McGee and Molly* was on the air for nearly a quarter of a century, finally ending in 1957. The television version debuted in the fall of 1959; it was a ratings disaster and was canceled after five months. Hal March's TV waterloo was considerably different. He emceed *The $64,000 Question* for three seasons, all in prime time with high ratings. But the quiz show cheating scandals of the late 50s forced his show off the air in November 1958.

Opposite Sara Berner, Sweeney played a character who, while dearly loved by her, was hopeless, hapless, and witless. In one adventure, they fled a building after they had discovered a corpse, and the next day…

SARA: The paper says witnesses at the crime scene saw a
            man and woman leaving the building.

MELVIN: I'll bet they did it! Those murderous villains
            deserve the gas chamber!

SARA: Honey...baby doll...that man and woman, that's us!

MELVIN: Sara, we're fugitives?

SARA: Yeah, ain't it fun?

In addition to trying to keep up with his imaginative sweetheart, who was rapidly changing her voice and disguises, Melvin had to endure the withering wrath of his grocery store manager, Mr. Sacks. This was the voice of Frank Nelson, who had achieved a measure of fame with one single, elongated word on Jack Benny's show, "Yeeeesssss?" which he emoted numerous times as the floorwalker, and etc. Fortunately for Melvin and Sara, Mr. Sacks was not that bright either. In one scene, where Sara found counterfeit money, she asked Sacks to go get a T-Man immediately. Sacks returned with a tea salesman from his grocery store.

Today's radio fans, listening to the two surviving episodes of this series, will be convinced of the range and variety of Berner's vocal skills. In the "Missing Book" episode, she began with an Edward G. Robinson impression, later switched to being a Greek gypsy fortune teller, then impersonated Mae West (while she passed herself off as Sassy Sara from Saratoga) and finally returned to her standard voice to join announcer, Frank Martin, in praising the delicious taste of Wheaties. The "Missing Necklace" episode contained more characters for her to create. She started with an impression of Lurene Tuttle's "Effie" from the Sam Spade show, switched to a nasal twang to sing "I Can Dream, Can't I?" changed to an Italian countess (and introduced Melvin as her American limo driver), followed by a clipped British dialect as "the Duchess of Windemeer," then a Chinese washer woman, and concluded as a raspy gangster's moll. To paraphrase the Bard, neither age nor custom could limit her infinite variety.

While some of the jokes seemed recycled from vaudeville routines, and the plot twists were frequently just plain silly, i.e. Melvin battling thugs, who had machine guns, by throwing bunches of radishes at them, the overall effect was still enjoyable for the audience.

The three leads stayed in show business, with small roles in television and film. Both Berner and Nelson, who were regulars on Jack Benny's radio series, also got roles on his television show. She was seen in a brief appearance in Alfred Hitchcock's *Rear Window*, but usually her film roles were limited to a voice part, as in *North by Northwest* and *Road to Morocco*. Berner died in Los Angeles in December 1969; she was only 57. Nelson passed away at the age of 75 in September 1986. Sweeney, who was younger than the other two, died in 1992 at the age of 73.

KITTY ARCHER

Kitty Archer was the "Mouse" on radio's *McGarry and His Mouse*. Mercifully, only three programs have survived from this abysmal series so that future generations will not have to waste much time on it. The series began on June 26, 1946 on NBC as the summer replacement for *The Eddie Cantor Show* and remained on the air until September 25, 1946. It was a sustaining series for the entire summer on NBC. Later, Mutual found a sponsor (General Foods), resurrected the series in January 1947 with a new cast, and aired it until March 31, 1947.

Despite the fact that it was a sustaining show, NBC had a full orchestra, under the direction of Peter Van Steeden, on stage with the cast and sound effects crew to impress the live audience. The orchestra played at least three full numbers in the course of this 30-minute program, usually pop tunes of the '40s or old standards, such as "I'm Always Chasing Rainbows" or "Who's Sorry Now?" These musical interludes were very unusual for a program in which the lead was a police detective. However, this was no ordinary detective show; it was a limp comedy in which the lead character worked in law enforcement.

The radio series evolved from fictional articles written by Matt Taylor for *This Week* magazine. The stories concerned the hapless adventures of an inept police detective named Daniel McGarry and his girlfriend, Kitty Archer, whom he nicknamed "Mouse." Taylor had little to do with the show, other than collect his royalties; the scripts were written by Milton J. Kramer. Kramer was a fine writer, as long as he was writing a serious crime drama, and he did well scripting *Counterspy, Famous Jury Trials,* and *Nick Carter*. But comedy was apparently not his forte, since the dialogue for *McGarry and His Mouse* was silly, simplistic, and forgettable.

There was no shortage of talent behind the microphone, even though there were a series of three different pairings of McGarry and his Mouse. The trio of actors who played McGarry (at different times, of course) were Roger Pryor, Wendell Corey, and Ted de Corsia, all successful and talented fellows who regularly appeared on network radio. The three women portraying Kitty Archer were likewise skilled professionals with regular roles in significant series: Shirley Mitchell, Patsy Campbell, and Peggy Conklin. Mitchell, in addition to Kitty, was Alice Darling on *Fib-*

ber McGee and Molly and the enduring love interest of *The Great Gildersleeve*, his Southern belle, Leila Ransom. Mitchell was also one of the four women who took her turn playing Claire Brooks on *Let George Do It*. Campbell had a long career in network soap operas; her many roles included Joan on *The House in the Country* and Patti on *Rosemary.*

Peggy Conklin (born Margaret Eleanor Conklin in Dobbs Ferry, New York) starred in soaps and comedies; she was the little sister on *Big Sister* and had a regular role on *That's My Pop*. Conklin had come to network radio after success on the Broadway staqe (she was the female lead in *The Petrified Forest* with Humphrey Bogart) and a Hollywood career of five routine movies. Conklin was about 40 years old when she played Kitty Archer, but obviously she sounded much younger.

But all the talent of these performers was to no avail in producing an enjoyable program. Each episode centered around Daniel McGarry, a bumbling police detective, his bright and perceptive fiancée, Kitty Archer, who worked at the airport, and his Uncle Matthew, who was also Dan's boss at police headquarters. The rest of the recurring roles on the show were characters who seemed to belong somewhere between *Duffy's Tavern* and any Warner Brothers cartoon. For, with the exception of Kitty, every one else in the program seemed to be infected with a permanent dose of stupidity.

Uncle Matthew, who spent most of his time shouting, was played first by Jerry Hartley, and then by Jerry Macy, both of whom tried for an Irish accent. Kitty's mother was the voice of Betty Garde, who was also the title lead in *Policewoman* during that same time. Thelma Ritter portrayed Bernice, or "Bernie," an associate of Kitty's. Bernie's boyfriend, Joe, played by Carl Eastman, was limited to two words at a time, the first of which had to be "yeah."

BERNIE:	I used to feel terrible when Joe left me to go to night school.
JOE:	Yeah, night school.
BERNIE:	But don't worry; they say the jails now are very nice.
JOE:	Yeah, nice.

Why that snappy dialogue would merit loud laughter is hard to imagine, but it did. The live audience, who can be heard on all three recordings remaining of this show, was obviously having a great time. Eastman, particularly, received hearty laughter on every one of his lines.

Although some of the episodes dealt with crime, this was not a requirement. One program filled its thirty minutes relating the problems of McGarry trying to help the police department win the upcoming glee club competition with other departments of the city. Trying to preserve McGarry's fine singing voice, Uncle Matthew takes him off night duty so he doesn't catch cold. Two firefighters hatch a plan to make him sick by getting him to stand in the draft in the firehouse while they spray the hose on his shoes. When this doesn't work, a third firefighter gives McGarry free tickets to a boxing match, which delights him, for in his words, "I love leather-pushers!" At the boxing arena, he gets excited and starts sweating, and in the seat behind him, a fourth firefighter is fanning him furiously, hoping the resultant draft will give McGarry a bad cold. But this devise fails too. However, the next day McGarry comes down with measles, and his main competitor in the Fire Department, while visiting him, also contracts the disease. Both are sent to the hospital while their frantic supervisors attempt to get them out in time for the glee club competition. They are both released at the same time, sing at the city-wide competition, but when the declared winner turns out to be a street cleaner, the police commissioner faints. The whole program involves a lot of screaming and scheming, with only Kitty Archer to calm things down. Not counting the three musical interludes, we have an entire half hour of pretty silly stuff, and not much of it very funny.

In another equally inane episode, McGarry is instructed by Kitty to bring his paycheck to the bank, where she will see to it that it is deposited correctly. However, at police headquarters, one of the squad rooms is filled with gambling paraphernalia, seized during a raid the previous evening. All the policemen are gambling and McGarry loses his entire $ 200 paycheck in this pseudo-Vegas atmosphere. He later meets Kitty at the bank and tells her only that he has "lost" his paycheck. She assumes it fell out of his pocket and she places a notice in the Want Ads, offering a reward of $50. Next, she gets a telephone call from a man who says he found the $ 200 and when he returns it to her, he gets his reward in a $ 50 check. That night, McGarry and Kitty spend several of the bills at different establishments. All of those businesses the next day report receiving counterfeit cash, and Uncle Matthew details McGarry to find those who passed the bogus bills. After Kitty finds out about the fake currency, she quickly concludes that she and McGarry must have passed these bills, after obtaining them from the phony Good Samaritan. She pulls McGarry aside:

DAN: Mouse, you're acting terrible mysterious.

KITTY: I know who's been passing out those counterfeit bills.

DAN: You do? Tell me, Mouse, and I'll put the arm on him!

KITTY: I don't think you will want to put the arm on him…it's somebody close to us.

DAN: (*Outraged*) Kitty! You don't suspect Uncle Matthew?

KITTY: No, it's not Uncle Matthew…it's you!

DAN: Hey, Mouse, this has got me bewildered. How come I show up a criminal? You mean I'm the man I been looking for?

KITTY: Yes, Dan, you are…

DAN: For goodness sake…this here's fierce. Do you know what this means? It means I gotta arrest myself!

Despite this uninspired nonsense, the series was popular with its live audience, as well as its thousands of radio listeners. At that time, Armed Forces Radio Services (AFRS) generally requested only popular shows to make airchecks for AFRS of programs done live so they could be rebroadcast to military abroad. Regular airchecks were made of *McGarry and his Mouse* under that arrangement.

JANE SHERLOCK

In the summer of 1946 a new lady sleuth series, *Meet Miss Sherlock,* was launched by CBS from the West Coast. Many listeners considered it to be a comedy detective program, but that was certainly not what its creator, E. Jack Neuman, had in mind. Nearly 50 years later, in an August 1994 interview, on NPR's *Weekend Edition,* which was hosted by Harriet Baskas, Neuman was heard explaining that his idea for Jane Sherlock was to combine innocence with determination, based upon character traits of a film actress he knew personally:

Meet Miss Sherlock, Monty Margetts (*Courtesy of Monty Margetts*)

NEUMAN: The trick was from an old friend of mine named Ida Lupino, whom I'd done movies with, and Ida always appeared as though she did not know what she was doing, that she was totally helpless and hopeless. Well, of course, she was totally in charge of everyone instantly and (she) knew exactly what she was doing. Well, I applied the same rule to writing fiction for radio.

In the scripts that Neuman and fellow author Don Thompson wrote, Miss Sherlock was an amateur crime-solver whose actual job was that of a buyer for Blossom's Department Store in New York City. Her fiancée, Peter Blossom, was a civil attorney and the son of the owner of the department store. Jane and Peter would stumble into a homicide and usually solve it before the local authorities, Captain Dingle of Homicide Division and his hapless assistant, Tollison. The standard opening for the show was:

PETER:	Ah…Jane?
JANE:	Yes, Peter?
PETER:	Now?
JANE:	Now what, Peter?
PETER:	Will you marry me now, tonight?
JANE:	Oh Peter, I'm so sorry; I can't tonight. Tonight I have to solve the Case of— [She named the episode]
PETER:	Ohhhhhhhhh.
MUSIC:	[Theme Song] "A Little Bit Independent"

The theme song, *A Little Bit Independent,* originally composed in 1935, had been recorded by many major vocal artists and was still popular in the 1940s. The piece was written by tunesmith, Joe Burke (1884-1950), with lyrics by Eddie Leslie. Both of them worked primarily on music for motion pictures and another one of their hits was *Moon Over Miami.* Burke had been composing for the early talkies with another lyricist, Al Dubin, and their popular songs included *Tip Toe Through The Tulips* and *Painting The Clouds With Sunshine.*

There were two separate versions of *Meet Miss Sherlock,* one in the summer of 1946 starring Sondra Gair, and a second one in the fall of 1947 with Monty Margetts in the title role. For the rest of the cast and crew, some confusion still exists. As might be expected, the memories of Gair (from an on-the-air interview in 1990) and Margetts (personal correspondence with this author in 1990 and 1996) differ somewhat. Neither of them jibes exactly with the cast credits on either of the two existing audio copies, both of which have Gair in the lead.

Gair, a native of Cicero, Illinois, graduated from Northwestern University in Chicago, with a major in theatre and a minor in journalism, and her drama classmates included Charlton Heston and Patricia Neal. By the time she had finished college, Gair had major roles in Chicago juvenile adventure shows, including *Jack Armstrong,* and several soap operas, one of which, *Masquerade,* was moving to Hollywood. Heston was outraged that Gair was headed for Hollywood with *Masquerade,* which he considered demeaning since they both planned to conquer the Broadway stage first. But Gair followed Horace Greeley's advice, not Heston's, and she was successful on West Coast network programs. She was delighted to discover that *Masquerade* added two members to the cast, Herbert Rawlinson and Francis X. Bushman, both favorites of hers from silent pictures. She was also in mystery dramas (*Suspense*), comedies (*Meet Corliss Archer*) and in July 1946, won the lead in *Meet Miss Sherlock.*

A 25-year-old, relatively unknown actor, William Conrad, was cast as Captain Dingle; he would later achieve great popularity on *Escape,* starting in 1947, and *Gunsmoke,* which debuted in 1951. The identity of the actor playing Peter Blossom in 1946 is not certain, but it was probably Jack Petruzzi, who had worked with Gair earlier in *Masquerade.* In the closing credits of the episode "Case of the Dead Man's Chest" (an aircheck from July 17, 1946) none of the cast is identified. However, announcer Tip Corning does mention the director, Dave Vail, the writer, Don Thompson, and the organist, Eddie Dunstedter. In a 1990 interview, Gair recalled a few of her supporting cast, including Paul Frees, Jeff Chandler (who was then using his real name of Ira Grossel) and David Ellis. This series ended in late September 1946.

About a year later, on September 28, 1947, Dave Vail and E. Jack Neuman resurrected *Meet Miss Sherlock.* Gair had left California to rejoin her husband in New York; he was recently discharged from the U.S. Army. Bill Conrad was busy with *Escape* and other network assignments so these two performers had to be replaced. Barney Phillips, who had played occasional roles in the 1946 series, was elevated to Conrad's place, doing the voice of Captain Dingle. Betty Moran was originally chosen to play the title lead, but after one episode, she was replaced. Monty Margetts, in correspondence with this author in 1990, recalled that Moran, a good friend of hers, "…evidently wasn't dithery enough to please…Vail, Thompson, and Neuman so they put me in and I stayed 'til the end."

Monty Margetts (that's her birth name) was born in 1912 and grew up in the Seattle, Washington region. In August 1929, when she was only 17, she got her first acting job at a Seattle radio station. After high school, she spent eight years doing stage plays at the Seattle Repertory Playhouse, many with Howard Duff, who was five years her junior. In the summer of 1937, she and Duff were cast in the leads of the theatre company's production of *Taming of the Shrew,* set to open that September. However, in July she and Duff went to San Francisco and there both found so much good-paying work in radio, they contacted the theatre in Seattle and advised they were not coming back.

Margetts stayed in San Francisco until 1945, doing most of her radio work at KPO, the NBC affiliate in the Bay City, including a stint at what she would later term, "the first female all-night disk jockey ." Radio studios were male dominated in those days, but NBC old-timers recall Margetts fondly as "one of the boys" who not only played poker with them, but also held her own in animated political discussions. Margetts also became friends with another radio performer, Natalie Park Masters (who would later portray *Candy Matson*) and the two women remained very close until Natalie's death in 1986.

Duff spent only a few seasons in San Francisco and then migrated south to greener radio pastures in Los Angeles, with regular appearances on *Screen Guild Players, Suspense,* and *Hollywood Star Time.* His soaring popularity as the lead in *The Adventures of Sam Spade,* debuting in 1946, led to a succession of movie roles beginning a year later. In 1945 Margetts moved to Los Angeles, upon being cast in NBC's *Woman's Secret* as "Mrs. John Taylor" in which she was the lead, the announcer, and also did all the commercials for General Mills.

For the 1947 version of *Meet Miss Sherlock* in which Margetts starred, the identity of the actor playing Peter Blossom is still unknown. Margetts remembered his first name was, coincidentally, Peter, but his last name escaped her. She said he later became a victim of groups which accused him of being a suspected Communist and he was driven out of show business. Murray Wagner was the program's announcer. Margetts thought that live music was provided by Milton Charles and his orchestra, however it seems unlikely that this unsponsored series would have been able to afford an orchestra. It's more probable that Charles merely played an organ, just as Eddie Dunstedter had in the 1946 version. Charles is remembered for his later work as organist on the juvenile adventure series, *Straight Arrow.*

Although only two copies of this half-hour show have survived (both from the 1946 series, as previously noted) and the scripts have not yet surfaced, researchers have uncovered eight episode titles which suggest the playful tone of *Meet Miss Sherlock:*

July 3, 1946:	The Case of the Scrubwoman Who Wasn't Mamie
July 10, 1946:	The Case of the Hamlet in a Straw Hat
July 17, 1946:	The Case of the Dead Man's Chest
August 8, 1946:	The Case of the Cock-Eyed Parrot

September 12, 1946:   The Case of Wilmer and the Widow
September 19, 1946:   The Case of the Pink Elephant
October 19, 1947:   The Case of the Talented Monkey
October 26, 1947:   The Case of the Society Murder

Listening to the two extant copies demonstrates not just the zesty humor of the program, but also the courage, logical reasoning, and tenacity of Jane Sherlock. In "Dead Man's Chest," Jane bought a rosewood chest at a local auction and her suspicions were triggered when two people separately tried to buy it from her at amounts greater than she had paid. Peter examined the chest and assured her it was empty, but she persisted in an examination and found a secret panel in the chest. The hidden contents consisted of a skeleton from a murder seven years prior. At the end of this mystery, the armed killer confronted Peter and Jane, but she kicked a chair at the murderer, allowing Peter to subdue him.

In the episode entitled "Wilmer and the Widow," the police had Jane's acquaintance, Wilmer, under suspicion for murder so she enlisted Peter to legally represent Wilmer. After this, a tough hoodlum threatened Peter, and when this arrogant thug was later killed, his mysterious lady associate, Yvonne, was the focus of the homicide investigation. Jane correctly deduced the hideout location of Yvonne, but when she and Peter went there, Yvonne knocked out Peter and almost escaped. Jane closed the case by unmasking the real killer, based upon the inscription of a ring the murderer had engraved.

A recurring joke, used at least once in every episode, involved Jane using her pet nickname for Captain Dingle, a term which had a slight sexual innuendo and also enraged the Homicide officer. A typical example would be Jane saying, "But the killer wouldn't know that, Dingey." Immediately Conrad, as Dingle, would bellow: "Jane, don't call me Dingey!"

After the second version of *Meet Miss Sherlock* ended in late October 1947, after a short run of five weeks, Margetts continued in radio, including commercials and voiceovers. But when her mother became seriously ill, the dutiful daughter gave up show business and went back home to become her mother's primary caregiver. At the time of her mother's death, Margetts had not worked for a few seasons. Jack Webb, who never forgot his old friends from San Francisco, cast her in his television show, *Dragnet,* and then he found her an agent. For the next thirty years, Margetts was a busy TV performer, usually in small parts in network shows, *Adam 12, Red Skelton Show, I Spy, My Three Sons,* and many more.

She had a major recurring role in the mid-1960s as Una Fields in ABC's *The Tycoon,* which starred Walter Brennan. Margetts also found a lot of film work at both Warner Brothers and Disney; she was seen in *On A Clear Day, Angel in My Pocket,* and *Any Wednesday.* "But one day in 1984," Margetts would later write, "I realized that I had been working for 55 years, so I retired myself." She enjoyed the next 13 years, busy with her hobbies and her grandchildren. Margetts died in February 1997; she was 85 years old.

Gair's years after *Meet Miss Sherlock* were much different from those of Margetts'. To please her husband, Gair gave up show business and concentrated on being a good wife and mother. She moved their family from New York to Los Angeles and then to Chicago, as her husband's jobs changed. In the late 1950s, her husband dumped her and ran off with her best friend, leaving Gair with their three young children, a daughter aged six, and two sons, one four years old and the other a six-month-old baby. Years later, in 1967, she gradually got back into radio, doing an interview program for NPR in San Francisco called *Mindscope.*

She remained in broadcasting, and seven years later, after a return to Chicago, worked for NPR at WBEZ, with on the air coverage of arts, news, and politics. In 1979, Gair originated a one hour daily program, *Live At Lawry's ,* which was filled with interviews of entertainment stars, European royalty, and sports figures. By the mid-1980s, she was heard on another show she created, analyzing foreign news; the program was entitled *Midweek With Sondra Gair.* She received many honors for her radio work, including induction into the Chicago Journalism Hall of Fame and "The Woman of Influence Award" presented by the National Association of Jewish Women. Gair passed away on May 25, 1994, a victim of breast cancer.

William Conrad had preceded Gair in death by only three months, on February 11, 1994. He had made the transition from radio to television effortlessly and starred in two lengthy series, *Cannon* in the 1970s and *Jake and the Fat Man* in the 1980s, as well as *Nero Wolfe,* which ran only seven months. He could be seen or heard, nearly all the time, from his movie roles to his vocal work on the popular TV cartoon, *Rocky and Bullwinkle.* Behind the scenes, he was a hardworking producer/director for Warner Brothers and was responsible for many projects, usually western movies. He was 73 when he died.

SUSAN BRIGHT

Although Susan Bright did not actually have her own show, she was a lady detective on radio. On *The Philip Morris Playhouse*, Bright solved crimes in 1942 in a continuing segment within this CBS half-hour variety program.

A brief historical review of this series, and its interior sections, should be provided. The Philip Morris Company began its variety show on network radio in February 1937 and over the next several years, changed the name, the format, the orchestra, and the performers. It was first called *Johnny Presents*, named, of course, for the Philip Morris spokesperson and living trademark, Johnny Roventini, a man 4 feet high, attired in the red coat, black pants, and pillbox hat of a hotel bellhop. His ringing cry, "Calllll for Philip Moooorr-rreees," over the musical strains of Ferde Grofe's "On The Trail" theme from his *Grand Canyon Suite*, may well be the most memorable advertising pitch in the last eight decades.

The name of the program was later changed to *Philip Morris Presents* and then it became *The Philip Morris Playhouse*, a name it retained through the late1940s. Its usual format was a 15-minute dramatic or comedy sketch, sandwiched between live orchestra music. The orchestras of Leo Reisman, Russ Morgan, and Ray Bloch were used in succession over the years and they played popular tunes of the day. The program was done with a live studio audience, to whom Philip Morris gave free souvenir programs for each performance, listing the songs, vocalists, and guest stars in the sketch. These souvenir programs contained astounding medical reports, such as:

> "As reported in leading medical journals, tests by doctors have proven conclusively that on changing to Philip Morris cigarettes, every case of irritation of the nose and throat, due to smoking, cleared completely or definitely improved. It's not only good judgment to smoke Philip Morris, it's good taste too."

Responsibilities changed in terms of who was in charge of the quarter-hour sketch in the middle of this variety show. Charles Martin originally directed the sketch, but was replaced in 1938 by Jack Johnstone, who then directed several episodes about a psychic detective and a second series involving heroes of the Royal Canadian Mounted Police. By early 1941, the dramatic sketch was replaced by a comedy segment and a movie star, Una Merkel, was hired as the lead. For most of that year, she starred in a humorous sketch entitled *Nancy Bacon, Reporting*, the happy adventures of a zany female reporter.

The radio section of the *New York Times* magazine on December 28, 1941 (exactly three weeks after the attack on Pearl Harbor) contained the following announcement:

> "After a spell as a dizzy reporter in *Nancy Bacon, Reporting*, Una Merkel will presently become a vertiginous sleuth in *Susan Bright, Detective*. Same time, same place: 8 o'clock, Tuesday nights on WEAF."

The remarkable thing about this brief announcement was not that it utilized the word "vertiginous," a little known synonym for dizzy, but that it contained no mention of *The Philip Morris Playhouse*. The cigarette company officials, and their advertising representatives, must have blanched at this item since it highlighted only Una Merkel and her two comedy segments, with not one word about the radio program that carried this sketch, or the company which sponsored it. After all, what was the point of naming a radio program after your company, if the newspapers weren't going to use that name? However, the item was a testament to the star power of Merkel that the public knew, without being told, what radio program featured her comedy segment.

No audio recording, or radio scripts, have yet been located for either *Nancy Bacon, Reporting* or *Susan Bright, Detective*. How long Susan Bright operated her detective agency, while part of *The Philip Morris Playhouse*, and what kind of cases she solved has also not been uncovered by vintage radio historians. There is a presumption that Bright lasted until March 1942, when Tallulah Bankhead came on the program so Merkel probably left. But, for those three (?) months of Una Merkel playing this lady sleuth, there is no doubt it was laughter, not law enforcement, that dominated her 15 minutes.

Una Merkel (1903-1986) was born in Covington, KY, and by her 20s, had found employment in the emerging silent motion picture business. Based on her athletic prowess and physical resemblance to Lillian Gish, she was frequently used by D. W. Griffith as a stand-in and double for Gish. She gradually moved from bit parts to significant roles in the film industry as it converted from silence to "the talkies" and in the period 1923 to 1935, Merkel appeared in over fifty movies. As trivia fans know, she played Sam Spade's secretary in the original *The Maltese Falcon* (1931) which starred Ricardo Cortez as the famous detective. From 1935 to 1950, she divided her time between the two coasts, performing in Broadway productions and appearing in Hollywood films. In that era, she was almost always used on the silver screen as a comedienne, the best friend of the leading lady. One of her brief roles was memorable; she battled Marlene Dietrich in the saloon brawl scene in 1939's *Destry Rides Again.*

Radio work was never the main focus of Merkel's acting career, but she did have some significant roles at the microphone. She was a recurring, funny character on both *The Great Gildersleeve* and *Texaco Star Theater.* She was also a regular on *The Bob Burns Show* ; Burns told his radio audience that Merkel had "one of them smiles you could pour on a waffle."

In the 50s and 60s, this talented lady was seen regularly on stage, on television, and in the movies. While most of her motion pictures were comedic fluff, i.e. *The Fuzzy Pink Nightgown, The Parent Trap,* and *The Kettles in the Ozarks,* she also was cast in difficult, dramatic roles and her performances resulted in significant awards. In 1956, Merkel won the Tony Award for her acclaimed stage role in *The Ponder Heart.* Five years later, in 1961, she was nominated for an Academy Award for Best Supporting Actress for her brilliant performance in the motion picture, *Summer and Smoke,* adapted from Tennessee Williams' play, and starring Geraldine Page (who was also nominated for an Oscar) and Laurence Harvey. Merkel lost the Oscar in a close vote to Rita Moreno in *West Side Story.*

While there are no surviving copies of any episodes of either *Susan Bright, Detective* or *Nancy Bacon, Reporting,* nor have the scripts been uncovered yet, we can rest assured that the playful personality of Una Merkel provided both her studio audience, and her radio listeners, with enjoyment and mirth.

When Merkel died at the age of 82 in January 1986, she left behind an enviable legacy of accomplishments in all areas of the entertainment world. She was one of the few silent movie performers who easily made the transition to the "talkies." She alternated between great roles on the dramatic stage, in radio, and on the movie and television screens. It was a feat that few of today's mega-stars would ever attempt.

More Than Just a Secretary

During the Golden Age of Radio, secretaries in both government and civilian jobs must have been somewhat envious of their counterparts on radio's crime dramas. It that era, most real life secretaries had fairly structured duties: take dictation in shorthand, type requested documents, answer the telephone, and file papers. Radio's secretary-sleuths occasionally engaged in such standard office tasks, but the majority of their time was spent examining crime scenes, discovering evidential clues, interrogating suspects, and helping their boss unmask the guilty.

CAROL CURTIS

The husband and wife team of Adele Jergens and Glenn Langan were the co-leads in the 1953 syndicated series, *Stand By For Crime*. Because it was syndicated, a total of 40 episodes, over half of those originally recorded, are still in existence today. The premise of this series was that a radio journalist, Chuck Morgan, used his news show, also called *Stand By For Crime*, at mythical station KOP in Los Angeles, to thwart crime and corruption. In these endeavors, he was ably assisted by his attractive and feisty secretary, Carol Curtis.

Neither Jergens nor her hubby had much radio experience prior to this program. Langan had the leading role of Barton Drake in a short-lived Mutual series, *Murder is My Hobby*, in the late 1940s, and its syndicated successor, *Mystery is My Hobby*. Jergens had virtually no microphone experience but both of them were comfortable and convincing in *Stand By For Crime*. Langan had a gritty baritone voice, akin to that of Howard Duff, which sounded great on radio.

We know very little about the rest of the cast and crew. OTR historian, Jim Cox, has determined that Bob Reichenbach was the show's producer and Howard McNear was occasionally in its supporting cast. As with most syndicated shows, cast members were not identified on the air.

Glenn Langen, Adele Jergens and Leif Erickson. (*Laura Wagner*)

Reichenbach had achieved some small amount of attention earlier in 1947 when he and Bill Rousseau helped turn Jean King into *Lonesome Gal,* the most successful, sexy female disk jockey in American radio history.

Both Jergens and Langan spent most of their acting careers trying to become major Hollywood film stars, but despite their looks and talent, they were usually relegated to B pictures. Jergens was born in Brooklyn in November 1917, although her publicity materials later would claim 1919 or 1922. Her dancing lessons as a little girl paid off by the time she was in her late teens, when she found employment as a Broadway chorus girl and model. Some sources indicate she was once one of the Radio City Rockettes. At the New York World's Fair of 1939, she won the "Miss World Fairest" contest, which not only improved her Broadway roles, but also made her popular as a pinup model.

In 1944, while an understudy for Gypsy Rose Lee in the Broadway musical, *Star and Garter,* Jergens got to take her place for two weeks when Lee was ill. A local talent scout saw Jergens in this show and talked Columbia into giving her a movie contract. Although she was an attractive brunette, the studio ordered her to bleach her hair blonde; she would

Adele Jergens *(Laura Wagner)*

remain a blonde for the rest of her Hollywood career.

Jergens got an occasional role in a significant motion picture, but most of the time she was cast as a floozy, stripper, or gang moll in forgettable B-films. Some of her typical movies were: *Girls in Prison, Slightly French, Aaron Slick from Punkin Creek,* and *The Mutineers.* In *Ladies of the Chorus,* Jergens played the mother of Marilyn Monroe, who was her junior by only nine years. The comedy skills of Jergens deserved better; she

Adele Jergens, "Carol Curtis" (*Laura Wagner*)

was used in one Abbott & Costello vehicle and also several Bowery Boys films for Monogram, including *Blues Busters* and *Blonde Dynamite*.

As she struggled for better roles in more prestigious films, the man she eventually would marry was having the identical career problems. Glenn Langan was born in Denver, Colorado and was only five months older than Jergens. He worked his way up from regional stock companies to the Broadway stage, where his performance in *A Kiss For Cinderella* in

1942 led to a movie contract with 20th Century-Fox. They cast him in major roles opposite leading ladies, Jeanne Crain in *Margie* and Gene Tierney in *Dragonwyck*, but the studio was unhappy with the audience response. After *Wing and a Prayer* in 1944, his screen roles got smaller and he ended up in B-pictures such as *Mutiny in Outer Space* and *Jungle Heat*. In 1949 he and Jergens were the co-lead lovers in *Treasure of Monte Cristo*. That relationship was replicated off-screen and they were married in 1951. Unlike most short-lived Hollywood marriages, this one would last just over four decades, ending only upon the death of Langan in 1991.

By 1953, when the opportunity for a syndicated radio show appeared, neither of them could have been satisfied with their progress in the motion picture industry. Nor would their radio jobs pay very much, but at least it may have offered some respite from minor roles in insignificant films.

There were only three main characters in *Stand By For Crime*. Chuck Morgan, who did two radio broadcasts a day, at 7 and 11 PM, spent most of his day solving local crimes and exposing community corruption. His secretary, Carol Curtis, whose primary duties included writing, not just typing, his scripts, usually accompanied him to crime scenes, assisted in interrogating witnesses and suspects, and occasionally rescued him from harm. The third character was John "Pappy" Mansfield, the owner and manager of the radio station; he was the catalyst for some of Morgan's sleuthing enterprises and he frequently threw himself into the investigation. The only other recurring character was that of Lt. Bill Meigs, but who portrayed him and Mansfield has not yet been determined.

The opening of each episode began with the announcer stating the title, *Stand By For Crime*, which was followed by a crescendo on the studio organ. Langan, as Morgan, would then briefly introduce the adventure of the evening. Each time it was different; below is a typical one:

MORGAN: Hi! I'm Chuck Morgan. Maybe you've heard some of my news commentaries over radio station KOP in Los Angeles. Yeah, that's right; I'm the guy that's always sticking his nose in somebody else's crimes. I'm the sucker for busting up some dirty racket and then blabbing into a microphone about it…in hopes that you folks out there will write your Congressman or something…

Like *Pat Novak*, Morgan had a predilection for getting physically clobbered, and in almost every episode, this radio journalist was knocked unconscious by assorted goons. In one episode, he confronted Bugs Spencer, an unsavory gambler, who instructed his two henchman to "show Morgan the door." The crime reporter was quickly knocked out, and since he was also the narrator of the program, he told his radio audience what happened as soon as he woke up.

> MORGAN: Bugs' boys had a peculiar way of showing me the door. I didn't see the door at all. I was unconscious when I went through it. I woke up in an alley behind the 9th Street building. The sun was beating down full force on my face and flies were buzzing around a cut above my left eye. I got on my knees and crawled to an outside faucet and soaked my head in the cold water.

The scriptwriters did their best to make Morgan speak in the manner of the hard-boiled private eye. He referred to San Quentin Penitentiary as "The Q" and he always called his automobile "my jalopy." Morgan was curt and insulting to his secretary, concealing the fact that he was in love with her, and frequently told her "Drop dead, Glamourpuss." In one episode, he asked her a question and she replied, "I'm thinking" and he snapped back, "With what?"

Carol Curtis gave as good as she got. A scrappy vixen when the situation called for it, she did not let her subordinate status stop her from admonishing her boss. Following another beating of Morgan, this time by two prizefighting pugs, he was back in the office where she administered first aid to his facial wounds.

> CURTIS: I hope this will teach you to take someone's advice once in a while. If you had let Bill and Pappy go down there with you—
>
> MORGAN: Shut up!
>
> CURTIS: I won't shut up. You're a newscaster, not a detective!

In another episode, this one involving a local dope ring, the following ensued after a woman came into Morgan's office to claim her brother was a heroin addict:

> MORGAN: It's not very likely. Anyway, John Burdeen, who's running for Councilman, has promised to clean up all the drug dealers and—
>
> CURTIS: Oh, for heaven's sake, stop! Stop pretending Burdeen's such a lily-white. You know very well he wouldn't clean his own teeth if it would get him votes.
>
> MORGAN: Well…
>
> CURTIS: The worst tragedy that has happened to this city in the past one hundred years is John Burdeen.
>
> MORGAN: You talk too much, Beautiful.

But despite all the banter, some of it apparently hostile, Morgan and Curtis were in love. While he did not share this feeling with her, he had no reservation about confessing it to his radio audience. During one case, in which he was preparing for an undercover role, he pretended to be anti-American so convincingly that Curtis angrily left his office, and he lamented to his listeners:

> MORGAN: I had lost the respect and love of the one person who's most important to me.

Most of the cases that Morgan and Curtis solved were murder cases or illegal drug deals or even drug-related murders. Other investigative matters concerned gangsters, smuggling, governmental corruption, and racketeering. In one adventure, Morgan, encouraged by Mansfield, took on "the Commies," penetrated a Red cell, and unmasked "Mr. Big," who had illegally obtained some sensitive documents relating to U.S. defense.

Generally, the episodes were scripted with crisp plots, unusual twists, and strong characterizations. The organist for the series provided excellent bridges, stings, and background music. Sound effects were skillfully produced and the overall quality of the series was very good.

Carol Curtis acquitted herself with talent, courage, and grit. Her diplomacy and patience in their interviews of witnesses and suspects compensated for Morgan's routine lack of tact. She was forthright, dedicated, savvy, and, if required, aggressive. In one of their few adventures outside Los Angeles, which was a drug smuggling case in a border town of Mexicali, Curtis was wounded in the arm by a smuggler who fled the scene. The next evening, when Morgan and his associates were fired upon by a hidden assailant, she conked the gunman on the head with a wrench.

After this radio series ended, Jergens and Langan returned to the movie business but found no improvement in their roles. After her appearance in another B-movie, *Runaway Daughters,* in 1956, Jergens retired from show business. Her husband continued a little longer, hoping that his leading role of Glenn Manning in 1957's *The Amazing Colossal Man* would bring him stardom. It did not. He went into the construction business, coming back to the cameras rarely, including a few appearances on the television series, *Hondo,* in the late 1960s.

Langan died in January 1991 at the age of 83. Jergens was just four days shy of her 85th birthday when she passed away in November 2002. Their only son, Tracy, had preceded her in death one year prior.

PAT

Although the scriptwriters did not choose to give Pat a last name in the radio program, *Manhunt,* she was an integral part of the plot in almost every episode. As the secretary to the head of a local police laboratory, Drew Stevens, she accompanied him to crime scenes, suggested possible motives, did background checks on suspects, and on occasion, obtained clues through illegal means.

Manhunt was a crime series, produced and syndicated by Frederic W. Ziv Company, which also distributed *Boston Blackie* and *Philo Vance.* The *Manhunt* series, consisting of 39 total episodes, was first recorded and released in the 1945-46 time frame, and thereafter aired at different periods by various stations. To date, 18 audio copies have surfaced and are in trading currency in the hobby.

The program's strengths far outweighed its few weaknesses. The title, although catchy, was a misnomer. In law enforcement, a manhunt refers to the search for an escaped fugitive or a person suspected or charged with a felony. This process never took place on *Manhunt* since the perpetrator was always arrested immediately upon the solution of the crime. Secondly, the scriptwriters did not know the difference between robbery and burglary. The midnight theft of cash from an unguarded safe was termed a robbery, but such a theft is actually a burglary. The third weakness in the series was the hokey introduction to every episode:

SOUND: (*Crash of cymbals*)
ANNOUNCER: No crime has been committed…yet!
No murder has been done…yet!
No manhunt has begun…yet!
SOUND: (*Crash of cymbals*)

But except for these minor flaws, the *Manhunt* series was excellent and it holds up well today.

The scripts were adroitly crafted, with a logical progression in deductive reasoning, from the initial crime to the ultimate solution, not an easy task within a 15-minute program framework. The crimes, nearly all of them homicides, were unconventional, baffling, and achieved through unusual causes of death. Several plots involved variations on ingenious locked room mysteries. Although the clues were fairly distributed through each episode, the solution to the case and identity of the killer was almost always a surprise to the listeners.

Every episode ended with a very clever tag line, usually uttered by one of the police personnel, but occasionally by the killer in cuffs. An example of the latter was heard at the end of the episode entitled "The Clue of the Legal Loophole" in which a woman was charged with killing her husband by poison, after she discovered he had cut her out of his will in favor of a younger woman. When the police found evidence of her guilt and took her into custody, she told them that she had anticipated her arrest and had taken poison herself so she could not be put in jail. Or, as she phrased it in the closing line, "Where there's a will, there's a way…out!"

Syndicated series had a reputation for hiring unknown actors at modest salaries, but this was not the case at Ziv productions. *Manhunt* starred three of radio's best performers: Larry Haines, Maurice Tarplin, and Vicki Vola. Haines, who played Drew Stevens on *Manhunt* , had a long successful career in network radio, with occasional work on soap operas, but more often on crime dramas. He was in the cast of *Big Town, Gang Busters, Man Behind the Gun, The Chase,* as well as *Cloak and Dagger,* and had the title leads in both *That Hammer Guy* and *Treasury Agent.*

Tarplin, who played Sgt. Bill Martin, head of Homicide on *Manhunt*, worked with Haines on three other radio dramas: *The Chase, Now Hear This,* and *Cloak and Dagger.* He was the voice of *The Mysterious Traveler* and portrayed Inspector Faraday on *Boston Blackie.* The female portion of the law enforcement trio on *Manhunt* was the voice of Vicki Vola, playing Pat, the secretary of Drew Stevens. Vola was equally successful on the soap operas (*Brenda Curtis, Manhattan Mother,* etc.) and crime adventure shows. She was not only in the regular cast of *The Fat Man* and The *Cisco Kid,* she was also the co-lead in both *Foreign Assignment* and *The Adventures of Christopher Wells.* In addition to being a police secretary on *Manhunt,* Vola was, at different times, the secretary of *Mr. District Attorney* and *Johnny Dollar.*

The interplay of the two police officers and the secretary provided humor and interest to their radio adventures. Martin was an old-fashioned, brusque, arrogant man who easily jumped to (incorrect) conclusions. He frequently traded insults with Stevens, a methodical, urbane crime-solver, who despite working in the police laboratory, almost never used any forensic science tools, preferring his instincts and logic to solve crime mysteries.

> DREW: Nice busting in, Bill. Didn't you see the sign on the door that says "Private"?
>
> BILL: I don't believe in signs on doors.
>
> DREW: I can believe that; the one on your door says "Detective."

Pat played gin rummy with her boss, between murder investigations, and occasionally flirted with him to no avail, as shown in this exchange from the episode "Clue of the Movie Murder."

> DREW: A whole week without a good murder!
>
> PAT: If you keep ignoring the fact that I'm alive, there'll be a murder, but you won't be alive to solve it.
>
> DREW: Pat, my pet, one more word from you and I'll rub a chemical from my laboratory on you and you'll disappear.

By invitation, Pat frequently accompanied Drew to homicide crime scenes and was helpful in the eventual solution of each mystery. In "The

Clue of the Masked Murderer," she directed Drew's attention to a children's playground set on the roof of a building which led to unlocking a murder mystery. Then, in "The Clue of the Movie Murder," Pat located the dead cat of a homicide victim, proving he was poisoned at his home, not at the movie theater where the body was found. After Drew had finished his interrogation of a patrolman in "The Clue of the Crimson Corpse," and had obtained no information of value, Pat asked a question that jogged the officer's memory. His answer to her query refocused the investigation, leading to the arrest of the killer. But not all of her assistance was in the realm of legality. During another homicide investigation, at Drew's instructions, she went to a lawyer's office and secretly purloined a legal document, which was instrumental in destroying a murderer's alibi.

Pat's only failure came to light in "The Clue of the Magazine Murder," when she was supposed to serve as Drew's armed back-up. The two were in a hospital room, expecting the arrival of a murder suspect. Drew loaned Pat his pistol and placed her in a chair, hidden behind a curtain, and instructed her to come out with the weapon and cover the suspect, when Drew signaled her by saying her name. But alas, when the murder suspect showed up an hour later, Pat had fallen asleep so Drew was unable to get her attention behind the curtain.

SUSPECT: Are those the notes you're holding? The notes on the Kirkland murder mystery?

DREW: Yes, but we're standing PAT on them…I said we're standing PAT!

SUSPECT: Hand 'em over…C'mon now!

DREW: I will as long as you don't PAT me over the head with that gun.

SUSPECT: That's better…hmmmm…All that trouble for nothing. There's nothing here.

The murder suspect bolted from the room and Drew discovered his back-up was sound asleep in the chair. Fortunately, the suspect did not escape since Drew had placed a powder on the floor which glowed under his flashlight, so he tracked the suspect back to his office and arrested him.

Haines, Tarplin, and Vola must have been proud of their work on *Manhunt,* although the unknown scriptwriter on the series deserved a lot of credit. All three of the performers continued in the entertainment business after the decline of dramatic radio, finding jobs in television, the movies, and doing commercial voice-overs.

Larry Haines was probably the most successful of the trio in their years following their radio careers. He was one of Oscar's poker buddies in the film, *The Odd Couple,* which starred Jack Lemmon and Walter Matthau. Haines also had numerous television credits and, in 1981, he won an Emmy for his long-running role on *Search For Tomorrow.* He briefly returned to the radio microphone as an actor on *CBS Radio Mystery Theater.* Tarplin died in May 1975; he was only 64. Vola was just shy of age 70 when she passed away in July 1985.

Sandra Lake

George B. Anderson gets credit for creating the detective series *The Crime Files of Flamond,* which began airing January 4, 1944. It was a part of Anderson's *Meister Brau Mysteries* at WGN, Chicago; the other three programs in this group were *Mystery House, Country Detective,* and *Mystery Playhouse.* Bereft of a first name, Flamond was a sleuth and psychologist whose secretary, Sandra Lake, was nearly an equal partner in their crime solutions. The series had three fairly successful runs, the first as a WGN regional show (1944-48), and then twice on Mutual (1953 and 1956-57). However, only nine total audio copies of this 30-minute show still exist; one from the WGN era and eight from Mutual's 1953 version.

The WGN series featured two of Chicago's busiest actors, Myron Wallace as Flamond and Patricia Dunlap as his secretary, Sandra Lake. Although Wallace originated the role, he went into military service and spent 22 months on active duty in the Pacific, so another actor took over. Some sources list Arthur Wyatt as Wallace's replacement, however a recent examination of Anderson's original scripts reflect it was William Everett Clarke.

Clarke, a well known drama coach at the time, changed his name a few times in radio. He went from William E. Clarke to William Everett to W. Everett Clarke, finally settling on Everett Clarke, by dropping his first name. He was in Chicago radio for years, usually in a supporting cast position, although he briefly held the title role in *The Whistler.* Clarke was on *World's Greatest Novels* and also was in *Destination Freedom,* the Negro civil rights anthology series that John Dunning accurately termed "one of the most powerful and important shows of its day."

Wallace, who was born May 9, 1918 in Brookline, Massachusetts,

Patricia Dunlap in *Bachelor's Children*. Rear: Olan Soule, Patricia Dunlap, Marie
Nelson, Charles Flynn. Front: Marjorie Hannan, Hugh Studebaker.
(*The Stumpf-Ohmart Collection*)

went off to college at the University of Michigan at Ann Arbor. While
there, he found work at a local radio station, but in the late 1930s, he had
moved to Detroit, joined WXYZ, and became the announcer on *The Green
Hornet.* The usual custom for radio performers in those days, who started
in Detroit or Cincinnati, was to go to Chicago since that city had the
most radio work in the Midwest. Wallace made the jump to the Windy
City in the early 1940s, and despite radio's specialization of microphone
duties, he was among the few men who was an announcer, a performer in
dramatic shows, and a news reporter, all at the same period.

While Patricia Dunlap was almost as busy at the microphone as
Wallace, her early background is not as well documented. One of her first
starring roles in radio was playing "Nada," the love interest of *Og, Son of
Fire,* a caveman adventure series for juvenile listeners in 1934-35. She
graduated to Chicago soap operas, and had regular roles on many, including *Bachelor's Children, This Day is Ours, Ma Perkins, Today's Children,*
and *Lonely Women.* Dunlap was regularly cast in juvenile adventures also,
playing "Betty" on *Jack Armstrong, The All American Boy,* and its succes-

sor, *Armstrong of the SBI*. She had major roles in some comedy shows, i.e. the wife of *Cousin Willie* and the sister of *That Brewster Boy*. Contemporary photographs of her and Wallace in 1947, posing for publicity shots for The *Crime Files of Flamond*, make her appear about five years older than Wallace. This was apparently the case, as she was working on *Og, Son of Fire* when Wallace was still in high school.

In 1947 Wallace was not only the voice of Flamond, he was on all four networks in some capacity. He was the host of *Famous Names*, the announcer on *Sky King*, the narrator on *Fact or Fiction*, the storyteller on Parade, a bit player on some soap operas, and a news correspondent on *NBC News*. Although he was usually called then by his nickname of "Mike," he did not change to that name until 1951 when he was hired at CBS in New York City.

One of the more entertaining stories about Wallace, during the time he was the announcer on *Sky King*, comes from the actor who played the title lead, Jack Lester. Years later, Lester related this story to popular radio writer, Jim Harmon, who incorporated it into his 1992 book, *Radio Mystery and Adventure and Its Appearances in Film, Television, and Other Media*. Lester said that Wallace was very involved in the stock market and was frequently on the telephone with his stock broker right up to the air time for *Sky King*, which caused some of the cast to worry he might miss his opening. One day, a volatile one for the market, Wallace spent so long on the telephone, he did not get to the studio by the time the show had to start. Therefore Lester, using a slightly higher pitch for the announcer, and a lower one for Sky, read both parts on the air. Lester explained he was worried since the sponsor was trying to keep expenses down and Lester was afraid they might tell him to continue doing both parts at his same salary. "Maybe Mike thought of the same thing," Lester concluded, "He was on time after that."

The WGN series ended in 1948, and about five years later, Mutual returned *The Crime Files of Flamond* to the airwaves in January 1953, but it only lasted six months. Everett Clarke was back in the lead and his secretary was played by Muriel Bremner. Our documented knowledge of her is less detailed than of his. Even the spelling of her name is not conclusive; some sources list her as "Bremmer" but the former is probably correct. Her radio credits were primarily Chicago soap operas, including The *Guiding Light* and *Road of Life*. She was with Clarke in *Destination Freedom* and they also worked together on *Comedy Playhouse*, a humorous anthology hosted by Jack La Frandre. Coincidentally, she and Wallace had been paired as a crime-solving duo in NBC's 1949 drama, *Crime On The Waterfront*, but the program was apparently canceled after just two episodes.

In its third version, and second time on Mutual, beginning in April 1956, *The Crime Files of Flamond* had Clarke and Bremner as the co-leads once more. The standard opening of most episodes of *The Crime Files of Flamond* usually went something similar to this:

MUSIC: (*Organ sting*)

ANNOUNCER: Card number 239 from the Crime Files of Flamond

MUSIC: (*Organ sting*)

FLAMOND: A new file card please, Miss Lake......

SOUND: (*Typewriter clicking in background*)

ANNOUNCER: Flamond! The most unusual detective in criminal history...Flamond! Famous psychologist and character analyst...Flamond! Who looks beyond tears, jealousy, and greed—to discover the reason why!

Each episode utilized a few recurring gimmicks. At the beginning of each show, after Flamond crisply ordered a new file card from Miss Lake, he began dictating the case results, and her flying fingers could be heard on the Underwood keyboard. However, for the rest of the program, he always called her "Sandra," no doubt leading some listeners to conclude he only addressed her as "Miss Lake" when they were alone in their private office. Incidentally, with the exception of the first minute of the show, and the last one, neither were ever in their office.

Every case received a new card number, issued in succession by Flamond, and the announcer would continue to mention that number at every scene break. The last two items in each adventure took place quickly, back in Flamond's office. The announcer would introduce the first one by saying, "And now, the Basic Clue!" and Flamond would oblige. Next Lake would reveal the name she had just created to summarize the case, which today's audio collectors use as the episode title. Of course, the reason this term was not spoken until the end of the show was that if it had appeared at the beginning of the show, the clue within it would have given away the mystery.

With an opening such as the above, listeners heard that Flamond was trained in psychology and was a sensitive sleuth trying to find out why, as he looked beyond "tears, jealousy, and greed." A lofty goal to be sure, but one that was seldom met in this program, which was little more than a routine detective show. Advertising man, turned scriptwriter, George Anderson, admitted this to *Radio Mirror* in 1947 when he casually described his technique for writing each episode.

> "I don't think much about plot outlines and solutions. I just type away. I figure if I haven't got the goods on somebody by the third-to-last page, the listeners won't either. In that case, the story's no good, and I start over."

In neither the WGN series or the two on Mutual did Flamond actually use any of his supposed psychology training, so his resolution of each crime mystery merely incorporated the same methods of most radio detectives of that era: surveillances, interrogating suspects, examining crime scenes, and playing a hunch. In fact, in some episodes, Sandra Lake did most of the actual investigative work, presumably at a lower hourly rate. In "The Case of the Chick That Killed," Lake went out by herself and located suspects which she brought to Flamond so they could both participate in the questioning. She cracked the case later by arranging for her two primary suspects, Marie and Carla, to confront one another in a showdown that eventually resulted in the guilty party confessing.

In another adventure, "The Case of the Suspicious Scream," Flamond's delicate touch of an experienced character analyst was, once again, nowhere to be found. He badgered witnesses and potential suspects with tough and threatening questions which bordered on arrogance. (Bear in mind, this was the voice of Everett Clarke, although Mike Wallace could have played it the same way.) Unfettered by routine social norms, Flamond rousted witnesses out of bed at 2 a.m. to question them.

While his secretary probably understood the legal implications of "breaking and entering," Flamond apparently did not; to enter an occupied residence next to a haunted house, he smashed a door window and then unfastened the lock. On another occasion, this "most unusual" detective made Miss Lake hide in the closet of the home of a suspected killer while Flamond went next door to search for clues. While agreeing to remain in the closet of a potential murderer could be classified as either naive foolhardiness or steely courage, what we know about Sandra Lake would certainly point to the latter trait. She was so calm in the face of obvious danger, such as being confronted by an armed thug, that no listener would dare doubt her bravery.

Concerning the sponsors of these three versions of *The Crime Files of Flamond*, the WGN one was sponsored by Meister Brau for most of its run. The initial Mutual series in 1953 was first sponsored by General Mills and then by Lever Brothers. When Mutual brought the program back in April 1956, they began airing it on a sustaining basis, but failed to attract any advertisers since the upstart television industry was siphoning off available sponsors' money. So Mutual pulled the plug in February 1957. But although the central character was not as unusual as the show's premise claimed, it did have three fairly successful lives, which is two more than most radio detective series could claim.

After the third series ended in 1957, there is little on record about the respective careers of Everett Clarke, Patricia Dunlap, or Muriel Bremner, but presumably they remained in their chosen careers despite the era's diminishing opportunities for employment. Clarke died in September 1980 at the age of 68, in a homicide that mirrored some of those on *The Crime Files of Flamond*; he was stabbed to death in his Chicago office. Wallace, as most people know, transitioned into television as easily as Superman divested himself of Clark Kent's attire. This television luminary has been a mainstay on a series of CBS programs , and, as of this writing, is still working in front of the camera at age 85 on 60 Minutes.

Rusty Fairfax

In April 1946, ABC, who should have known better, began airing a substandard imitation of the Mutual Network's *The Crime Files of Flamond*. ABC called their counterfeit *Danger, Dr. Danfield*, and for no good reason, kept it on the air for a full year before discontinuing it in April 1947. But alas, this was not the end of this copy cat program; Teleways, a transcription company in Hollywood, CA, bought the series, recorded more episodes, and actively maintained it in syndication through 1951.

Flamond was a detective with a background in psychology and interested more in the "why" of crime, not just the "how." He began, and ended, every episode by dictating to his secretary, Sandra Lake, who accompanied him on every investigation. Danfield was a detective with a background in psychology and interested more in the "why" of crime, not just the "how." He began, and ended, every episode by dictating to his secretary, Rusty Fairfax, who accompanied him on every investigation.

Teleways, despite their mediocre programs in circulation, unabashedly promoted them all as though they were shows of network quality. They seemed to specialize in knockoffs. Teleways copied WLW's popular *Moon River*, which blended music and poetry, with their own diluted version called *Moon Dreams*. To lure fans of *National Barn Dance* and WWVA's *Jamboree*, this syndication company offered a show they thought was similar, *Barnyard Jamboree* (with "Round Boy" Jeffries and the Milk Maid Quartet). In the October 1948 issue of *Sponsor*, the primary periodical for radio station owners and managers, Teleways had a full page advertisement, which lavishly praised their current transcribed programs, and it described the Danfield series as follows:

> "DANGER, DR. DANFIELD 26 Fascinating 1/2 Hour Mystery Shows. Starring Michael (Steven) Dunn, Twentieth Century Fox & Columbia Star, as Dr. Danfield, Crime Psychologist. Brilliantly conceived and written by Richard Hill Wilkinson."

Dunne was billed as a "star" with both 20th Century-Fox and Columbia Pictures, appearing in many films as Stephen/Steve Dunn (Columbia) or Michael Dunne (Fox), such as *Once Upon a Honeymoon* (1942), *Junior Miss* (1945) and *Cha Cha Boom Boom* (1956).

The complimentary summary of the talents of Richard Hill Wilkinson in this advertisement is nonsense. This series, in both its concept and writing, was unimaginative, mediocre, and pedestrian. Incidentally, John Dunning in his 1998 vintage radio encyclopedia, *On the Air*, inadvertently lists the creator and writer of the series as Ralph Wilkinson, and subsequent radio reference books have repeated this error.

Neither Dunne, nor JoAnne Johnson, who played Danfield's secretary, Rusty Fairfax, had any other documented radio credits. If that's true, they may have wanted to drop this second-rate series from their respective resumes. The only other character, who appeared in nearly every episode, was Captain Otis of the local police, but the actor who played him is not recorded. The standard opening for each program, while somewhat more ostentatious than that of Flamond, nevertheless established that their training and tactics were identical:

> DANFIELD: The human mind is like a cave, beyond the light, there are dark passages and mysterious recesses. I, Dr. Daniel Danfield, have explored those unknown

retreats and know their secrets.

SOUND: (*Organ sting*)

ANNOUNCER: Dr. Daniel Danfield, authority on crime
            psychology, has an unhappy faculty for getting
            himself mixed up in hazardous predicaments,
            because of his astonishing revelations, regarding the
            workings of the criminal mind.

While Wilkinson sprinkled his scripts with adjectives describing Rusty
Fairfax (attractive, young, pretty) he refrained from using the term that
would characterize her best: annoying. In every episode, she was an offen-
sive, snippy, and vexatious pest. She criticized her boss, their clients, potential witnesses, and police officers, until Danfield had to tell her to pipe
down. Below is a typical exchange, which took place outside the hideout of
some bank robbers, where it was suspected that Patrolman Richards was
held hostage:

DANFIELD: So you assume whoever was in the house prevented
          Richards from leaving?

OTIS: That's exactly what we think, Doc.

FAIRFAX: Well, why don't you go up there and find out, for
           heaven's sake?

OTIS: For a very definite reason, Miss Fairfax, Suppose
           the people in that house are the bank robbers, and
           suppose they've taken Richards hostage—

FAIRFAX: Suppose they are. They can't hold him forever.
           Whadda you gonna do? Just sit around and wait.

DANFIELD: Miss Fairfax, may I make a suggestion?

FAIRFAX: Sure, Dan, go ahead…

DANFIELD: Stop asking so many questions. Let Captain Otis
          finish his story.

In her defense, one should note that crusty Rusty appeared to have
Danfield's best interests at heart. She was protective of him, cautioned him
not to skip his paid lectures to work gratis on a new case, tried to keep him
out of harm's way, and promoted his professional status. But she remained

arrogant, nasty, and obstinate in her opinions, which were delivered in a perpetual whine by JoAnne Johnson that still galls listeners today.

Because of the corny, and frequently hackneyed, scripts neither the doctor nor his secretary had much chance to shine. In one episode, entitled "Murder of Norman Mills," Rusty warned Danfield not to examine the crime scene after dark, but he ignored her caution, so she accompanied him around the estate grounds. When he decided to climb a tree to examine a window lock, she pointed out that the limb would break under his weight. It did, and the reckless sleuth was knocked out by the resultant fall. However, Rusty did not say "I told you so" since she was also unconscious on the ground, felled by a blow to her noggin by the unknown murderer.

To protect Danfield and Fairfax from danger, and rescue them from physical tribulation, author Wilkinson used another unrealistic plot device: the appearance of a character named Mario Contelessi. He was played by an unknown actor, who mimicked the grating Italian accent mumbled by J. Carrol Naish, the lead in *Life with Luigi*. Mario was an all-purpose protector who appeared on an average of every third episode. His strength and quick fists overpowered any adversaries of Danfield or his one-woman staff. Mario's relationship with the sleuths was not clear in the scripts so he may have been an "on-call" bodyguard.

Rusty, despite her constant harping, did contribute toward almost every investigative solution. During a "locked room" suicide inquiry, she astutely pointed out that since a key piece of evidence within the crime scene room had been wiped clean of fingerprints, it was unlikely that the victim's death was self-inflicted. In other episodes she noted inconsistencies in witness testimony and poked holes in suspects' alibis. Occasionally she made more sense than Danfield, whose conversation tended to be a trifle stuffy and, sometimes, pompous. The doctor had a habit of tossing off silly bromides:

DANFIELD: It is the common, everyday citizen, who upon
reverting to the criminal type, displays more
cunning than the seasoned professional.

Twenty-six episodes of this mundane series have survived, and in most of them, the plot stretched credulity. One example may suffice. "The Case of Oliver Norton" began when a theatre manager was blackmailed, by the actress he was sleeping with, into putting her in the lead role of his current play. After a harsh exchange of words, he strangled her and then carried her body to another office where he put her near an open gas jet. That evening he took the corpse to her apartment, sealed doors and windows, then opened the gas jets to fake a suicide. He accomplished all of this without any theatre personnel or neighbors noticing anything.

Next morning Captain Otis requested Danfield's help at the victim's apartment so Danfield and Fairfax obliged. That night, Danfield went to the manager's office to look for clues while Fairfax stalled the manager at his residence. Another actress burst into the manager's office, and shot Danfield in the head, thinking he was the manager. This second actress then drove to her apartment and committed suicide, but tried to make it look like murder. Danfield recovered from his wound, and finally solved the case, but how many listeners cared?

No matter what had transpired in the course of each investigation, most episodes ended in the same manner. Danfield, apparently forgetting how obnoxious Rusty had been, closed the case in his office by giving her a lovely compliment and a big kiss.

CLAIRE BROOKS

Although the Mutual broadcasting system kept *Let George Do It* as just a West Coast regional show for over ten years, it certainly deserved to be heard by the rest of the country. Since it was a transcribed series on the Don Lee Network, nearly 200 of these half-hour episodes have survived so virtually anyone today can still enjoy the strengths of this excellent program.

George Valentine was the leading character, and akin to Dan Holiday of *Box 13*, mystery, adventure, and excitement began with a request for help. Valentine's classified ad was slightly different from Holiday's "Adventure wanted. Will go anywhere, do anything" since the former's ad read: "Danger is my stock in trade. If the job's too tough for you to handle, you've got a job for me."

Another difference between these two shows was that Holiday's associate, Suzy, limited her assistance to her boss to picking up his mail and taking telephone messages in the office. Valentine's associate, Claire Brooks, accompanied him on every mystery hunt and shared the surveillance work, interrogations, and crime-solving. "Brooksie," as Valentine usually called her, was a lady of initiative, courage, and foresight, whose efforts made her his partner, in all but name.

Robert "Bob" Bailey was the voice of George Valentine from the series debut in November 1946 until about 1954 when Olan Soulé took over for the last year the show was on the air. At least four women played Claire Brooks: Lillian Buyeff was probably the first. Frances Robinson took over about 1949 and Shirley Mitchell had the role briefly, probably after Robinson although the documentation is not clear. For the last four years, Virginia Gregg portrayed Brooks.

Bailey, who was born in 1913, supported himself in a variety of non-acting jobs until he finally got a contract with Twentieth Century-Fox , about the same time his radio roles began to improve. He did some network comedies (*Mortimer Gooch* and *That Brewster Boy*) and a few soap operas (*Today's Children* and *The Story of Holly Sloan*). But his reputation in radio broadcasting was assured with leads in *Let George Do It*, and later, *Yours Truly, Johnny Dollar*.

Lillian Buyeff, who specialized in playing savvy dames, was usually cast in adventure shows, including *Crime Classics, The Green Lama*, and *I Love Adventure*, but playing Claire Brooks was one of her few co-leads. Both Frances Robinson and Virginia Gregg, who later played Brooks, also had experience as the sidekick to a radio detective. Both played Helen, the girlfriend of *Richard Diamond, Private Detective*, although at different times, of course. Robinson portrayed *Philo Vance*'s associate while Gregg was one of seven women who voiced Nikki on *The Adventures of Ellery Queen*.

When *Let George Do It* began, Valentine and Brooks shared the microphone with three other regular characters; Sonny, the kid brother of Brooks and their office boy (Eddie Firestone, Jr.), Caleb the elevator operator (Joseph Kearns) and Lt. Riley (Wally Maher). Within three years, all three characters were eliminated from the show. From 1949 on, the only other recurring character was Lt. Johnson of Homicide, played by Ken Christy.

The directors, first Don Clark, and later Kenneth Webb, consistently picked the cream of the crop from West Coast talent in their supporting cast. Among the veteran performers regularly heard in small roles were: Harry Bartell, Peter Leeds, Lurene Tuttle, Parley Baer, Alice Reinheart, Byron Kane, William "Bill" Conrad, Ted de Corsia, Lawrence "Larry" Dobkin, Lee Patrick, Ted Osborne, and Charlotte Lawrence. But the core strength of any radio series stems from its scripts, and the writing on *Let George Do It* was excellent. Most of the writing responsibilities fell to Jackson Gillis and David Victor, although there were occasional scripts by Polly Hopkins and Lloyd London. All of these authors consistently furnished strong plots, crisp dialogue, and logical resolutions to each crime mystery. Maintaining a high level of mystery and excitement in a crime drama over ten years is a difficult task, but this writing team fulfilled their mission.

Although the series had a standard format, the variations within it seemed endless, so listener interest rarely lagged. Either a letter, telegram, or telephone call for assistance set each plot in motion. A multiplicity of problems or potential crimes commanded the attention of Valentine and Brooks. They received requests from a woman who had premonitions of specific murders, an organizer of a magicians' convention who suspected deadly trickery, a young woman being driven mad by her domineering sister, and an art collector whose Korean statue attracted threats from unknown persons.

Valentine and Brooks were a tag-team of mystery-solvers who shared all necessary investigative duties. In a June 1944 episode, she even drove his automobile on a surveillance. While this may not seem startling in present day, prior to the 1960s no man would permit a woman to drive his car, unless both of his arms were broken. Brooks was frequently on her own, dispatched to interrogate suspects, gather evidence, or identify subjects. Her courage and physical ability were clearly demonstrated in an April 19, 1948 adventure, "Penthouse Roof," when she climbed from one high-rise terrace to another and then interrupted a confrontation between Valentine and a distraught woman who held him at gunpoint. Later, in that same episode, Valentine assigned her to check all drugstores in a twenty block radius, and her persistence paid off when she located the prescription for the poison which identified the killer.

Their relationship was all business when they were working a case; she called him "Darling" and he referred to her as "Angel" only in private moments. Otherwise, they addressed each other as "George" and "Brooksie." Although Valentine introduced his constant companion to others as "Miss Brooks." or less frequently, as "Miss Claire Brooks," he never used her first name in their conversations together.

As 1950 neared, Miss Brooks, perhaps weary of a semi-romantic relationship for nearly a decade, fell into a habit of ending each episode with a veiled proposal for a more significant commitment from Mr. Valentine. At the conclusion of the April 26, 1948 program, "The Wolf Pack," she was looking down from an apartment window at teenaged lovers, and called Valentine's attention to them:

BROOKS: Look down there, out of the window. It's Eddie and Emily sitting on the stoop.

VALENTINE: Oh yeah (chuckles). Offhand, I'd call that romantic, Angel.

BROOKS: And offhand, I remember a saying, "Speak for yourself, John."

VALENTINE: Hmmm?

BROOKS: If you know what I mean…

SOUND: (*Musical bridge*)

We never find out if Valentine knew the significance of that line from the 1858 narrative poem of Henry Wadsworth Longfellow, *The Courtship of Miles Standish*, in which Standish sends his friend, John Alden, to woo Priscilla Mullins in his behalf.

In another episode, which involved hypnotic suggestion to cover a crime, Brooks hinted that she might use hypnosis to get Valentine to pop the question. Later, another adventure concluded with a lady victim being saved by the two sleuths in time for her marriage. Brooks assured the prospective bride that they would attend her wedding and requested the bridal bouquet be thrown at her. Despite all the hints, Valentine remained cordially non-committed. Brooks was apparently resigned to her role as a crime-solver; in the January 21, 1952 program, "A Matter Of Honor," Valentine characterized her as a souvenir collector since she still had her childhood doll and her first dance program. Brooks responded by saying that her sentimental relics included some recent additions: a blackjack, an old faded poison pen letter, and a sawed-off shotgun.

Let George Do It finally came to an end in January 1955, and a few months later, Bailey took over the title lead in *Yours Truly, Johnny Dollar*, a role he would play for the next five years. He retired from show business in the mid-1960s and died in 1983. Little has been documented on the respective careers of Frances Robinson and Lillian Buyeff after *Let George Do It* went off the air. Virginia Gregg went on working, in both television and film, for several years. She appeared on many television episodes of *Gunsmoke* and had motion picture roles in *I'll Cry Tomorrow* (1955), *Spencer's Mountain* (1963), and *Guess Who's Coming to Dinner* (1967). Gregg also did voiceover work on many television cartoons until she retired about 1982. She died in 1986 at the age of 71.

And The Winner Is...

CANDY MATSON

The best radio series featuring a lady detective was undoubtedly *Candy Matson, Yukon 2-8209.*

It had all the strengths of a great radio drama: strong and likable characters, superb writing, authentic settings, fast-paced excitement, subtle humor, and high broadcast standards. This series was the product of a talented married couple, Monty Masters, its creator, writer, and producer, and Natalie Park Masters, who starred in the title role.

These two had been active in radio on the West Coast for many years. Natalie was born November 23, 1915 in San Francisco and grew up there; she was a second cousin of Celeste Holm. There were only eleven years between Natalie's stage debut at age seven, and her first radio job, a Carnation Milk commercial, when she was eighteen. Her uncle, Nathaniel "Jack" Thomas, who was the director of the Wayfarers Civic Repertory Theatre where they both worked, begged her not to go into radio: "You'll be bastardizing your art as an actress." This was in 1933, and two years later, she was appearing in several radio shows in San Francisco, including a soap opera, *Hawthorne House.* "This program was described by another cast member, Monty Mohn, as "sort of a poor *One Man's Family,*" but it ran for over a decade. Mohn, who was three years older than Natalie, changed his last name to Masters, and later did the same to Natalie by marrying her. It was her first marriage and his second.

Monty was more interested in writing and producing, though he continued to accept microphone jobs. He kept trying to create a radio series that would be best for the talents of him and his wife. In 1947 he got NBC to air his newly formed comedy as the summer replacement for *Truth or Consequences;* it was called *Those Mad Masters* and was done before a live audience. Using their real names, Monty and Natalie played a goofy married couple, whose confusing antics were supposed to be funny, and sometimes they were. But it was a fairly routine situation comedy and was not renewed; it went off the air in August 1947 after running two months. Two audio copies of it have apparently survived, but only one, "The Typewriting Championship," is in general circulation.

In the spring of 1949, Monty was at work on yet another new series, this time an adventure series about a San Francisco detective, Candy Matson. At this time, the Masters were close friends and radio associates of Jack Webb, and his primary writer, Richard "Dick" Breen. The latter two were responsible for the success of *Pat Novak For Hire*, which had debuted in San Francisco on radio station KGO in 1946, and was then doing very well on ABC's West Coast network. Webb as Novak, a hard-boiled maverick, solved criminal cases with the help of his alcoholic pal, Jocko Madigan. The character of Candy Matson resembled Novak, mouthing sarcastic quips, and Matson's boozin' buddy was Rembrandt Watson.

While the *Candy Matson* series was in the development stage, Monty was planning to play the title role himself. He and Natalie read a proposed script to her mother, who suggested it would be a better vehicle with her daughter playing the lead. Monty agreed and rewrote the part for a woman, keeping the same name for the private investigator.

On April 4, 1949 an audition disk was recorded under the working title, *Candy Matson, EXbrook 2-9994*. Casting the show was relatively easy for Monty and Natalie; they knew all the strengths and weaknesses of most radio performers in the Bay area. The role of Mallard, a police detective, was given to Henry Leff, a talented actor and longtime friend of theirs; Leff had had a major role in *Those Mad Masters*. Natalie talked Monty into casting her uncle, Jack Thomas, as Rembrandt Watson. It was an unusual choice for although Thomas was an accomplished stage actor, he had absolutely no radio experience, and sixteen years earlier, he had tried to dissuade Natalie against a radio career. The rest of the supporting cast, some of whom the Masters had worked with in *Hawthorne House* included Harry Bechtel, Lu Tobin (his name is misspelled "Lou" in most radio reference books), Helen Kleeb, Jack Cahill, Mary Milford and Harold "Hal" Burdick.

The series theme song, "Candy," was chosen by Monty after he changed the gender of this San Francisco private eye. This tune, written by Mack David, Joan Whitney, and Art Kramer in 1944, had become very popular in the ensuing years with successive recordings by several artists, among them, Johnny Mercer, Harry James, Nat King Cole, and Sammy Kaye. While it, as played by organist Eloise Rowan, was perfect for *Candy Matson*, the song was not limited to this crime drama . *True Detective Mysteries* , sponsored from 1946 to 1953 by the Curtis Candy Company, advertising their O'Henry candy bars, also used "Candy" as its closing theme song during that period.

NBC gave the series a green light and scheduled its network debut for June 30, 1949. Although audio copies today exist for the April audition and the second show of the series on July 7, 1949, there is no surviving copy of the premier episode. It used the same script as the audition, "The Donna Dunham Case," but some minor changes had been made in the intervening three months. Candy's telephone number in the title was changed from EXbrook 2-9994 to YUkon 2-8209. Mallard, the police detective who had only a last name in the audition, was accorded a first name, Ray, and he was promoted to Lieutenant. A more significant change modification involved the relationship of Candy and Mallard. Henry Leff, in a presentation to the Broadcast Legends in September 2002, recalled:

> "Regarding the arrangement with the character of Mallard, when we did the original, I played Mallard as a kind of avuncular, older man on the police force, trying to help this young private eye. When the series came back approved, we sat down and talked and Monty said, 'You know, maybe we ought to make them a romantic team' and I said 'fine'."

Monty then rewrote the script to indicate a romantic interest between Candy and Mallard so Leff played his role differently. For the Mallard in the audition piece, he used a slower, deeper voice, but then changed to a more energetic and slightly higher tone when he became Candy's beau.

The show was done live, even though the NBC engineer, Clarence Stevens, under instructions, recorded an air check of every episode. Monty Masters, who owned the series outright, had planned to syndicate the series later in other markets, including Canada, but these never materialized. All 94 of the original broadcasts were, at one time, on disk but over the years most were lost or destroyed. To date, only a dozen of the original shows have been uncovered, plus the audition show and a reprise, recorded in Los Angeles on September 21, 1952, under the slightly amended title of *Candy Matson, YUkon 3-8309.*

All of the rehearsals and broadcasts were done at the Radio City Building at the corner of O'Farrell and Taylor Streets in downtown San Francisco. This magnificent structure was built by NBC...by mistake! In 1940 that network had authorized the construction of a new edifice to replace its old facility at 111 Sutter Street, for at that time, NBC planned equal staffing in their two primary West Coast studios, Hollywood and San Francisco. (Incidentally, Dashiell Hammett placed the fictional office of Sam Spade in that same building on Sutter.)

By 1942, when the new building was formally dedicated, NBC had already transferred most of their Bay personnel to Hollywood. Virtually all of NBC network programming on the West Coast was now concentrated in the Los Angeles area, with only a much smaller presence in San Francisco. So this beautiful structure, with its three-story mosaic mural designed by C. J. Fitzgerald, was San Francisco's last gasp as a radio center. However, at least through the rest of the 40s, despite smaller markets, the building remained relatively busy with full day programming with most of its studios regularly in use, and both NBC and ABC (who also occupied the building) employed full orchestras.

When Monty Masters brought his *Candy Matson* crew into Radio City in 1949, there were seven studios of various sizes on the second floor, shared by NBC and ABC. Masters was usually assigned either Studio B or Studio C for *Candy Matson*, which initially had a small audience. As the series increased in popularity, and more people came to watch the program, it was moved to Studio A, which had the most space and could accommodate a few hundred audience members on folding chairs. Incidentally, about twenty years later, Studio A was featured in the motion picture, *The Candidate*, which starred Robert Redford. In the film, Redford runs into a supposed television building (actually, Radio City) and he goes up to Studio A on the second floor for an interior scene.

Although Monty and Natalie did not get to choose their production personnel, as they did their cast members, they must have been very pleased with the talented staff that NBC assigned to their program. Clarence Stevens was a first class engineer. Depending on their complexity, sound effects were performed by one, two, or on rare occasions, three, men. Bill Brownell was the senior sound effects artist and his usual assistant was newcomer, Julian "Jay" Rendon, only twenty-one years old. If a third person was required, it was usually Juan Transvita. In addition to the standard items on a soundman's truck, they had devices to imitate a cable car, fog horn, buoys, and various pier sounds.

The musician for the entire run of the series, Eloise Rowan, had played studio organ for many years on Chicago soap operas before relocating to San Francisco, where she was on several network and regional programs. In 2003 Henry Leff recalled her with great admiration: "She was a musical marvel...brilliant in her selection of book music and an inspired improvisational organist!" During one episode of *Candy Matson,* Mallard advised the lady sleuth that she might get arrested after she and Rembrandt assaulted a suspect. Candy responded "Yipes! I wonder how I'll look in stripes?" and Eloise quickly played "The Prisoner's Lament" ("If I had the wings of an angel, over these prison walls I would fly.") In another adventure the police lieutenant tried to delay Candy's exit from her apartment by suggesting they have a drink first. Candy retorted "From now on, Mallard, you'll have to earn your cocktails" and Eloise followed immediately with a few bars from "Cocktails For Two."

When the series began in June 1949, the announcer was Joe Gillespie, a solid, articulate man at the microphone. He held the job through that August and was replaced in September 1949 by Dudley Manlove, who remained the announcer for the rest of the run. Manlove's surname, coincidentally, indicated his sexual orientation, but his homosexuality was certainly no impediment to success in the entertainment industry. His biographer, Dr. Robert Kiss, who is a university professor in Coventry, England, recently advised that Dudley Devere Manlove, born June 11, 1914, was a child vaudeville star. He had long blonde curls and versatility on the stage as a dancer, singer, and comedian who also played the piano and the accordion. Manlove's stage appearances led to small roles in silent movies but his musical career ended about the time of his high school graduation, when a hit-and-run accident in Oakland, CA nearly killed him. After long hospitalization, he sought less strenuous employment in radio, beginning at station KLX. There, he invented the craze of "snap dancing," clicking his fingers near the microphone with such dexterity, he could imitate any famous tap dancer over the airwaves. This unusual technique took him all the way to New York City radio before the craze died out and he returned to California radio, working up to NBC staff announcer in 1944.

Henry Leff arrived on the West Coast shortly after Manlove started with NBC. Leff was born in Manhattan on August 20, 1918 and after high school there, got a degree in speech from Brooklyn College (1939) and his Masters in theatre from Cornell (1941). For the next four years he was in the U.S. Army and after his discharge, he and his wife, Sylvia, (whose professional name was Bobby Lyons) relocated to San Francisco where they both found jobs in radio. Leff also taught oral communication at City College of San Francisco, and within a few years, was heading up their broadcasting department.

Candy Matson, YUkon 2-8209, a half-hour drama, was aired on Thursday evenings for the first seven weeks but beginning with the August 22, 1949 broadcast, NBC moved it to Monday nights, and it retained that time slot until the series ended in May 1951. The cast and crew were remarkably consistent throughout the two years it was on the airwaves. Engineer Clarence Stevens was in the booth for almost every episode, except for a couple when Phil Ryder or Frank Baron filled in for him. Dudley Manlove was the announcer on virtually every program from September 1949 on, missing only two performances when his replacements were staffers John Grover and Phil Walker.

Monty Masters wrote the vast majority of the scripts, although he had some help, not always credited, in the second year. One of Leff's college students, teenaged Jerry Zinnaman, was trying to break into radio as an actor, but his inexperience limited his prospects. Zinnaman was a good writer and Monty accepted some of his scripts for *Candy Matson;* in lieu of a writer's fee, Monty would cast Zinnaman in the shows he wrote. At that time, supporting players on San Francisco radio were paid $ 15 per show. Although Monty produced and directed most of the episodes, he did turn the program over a few times to a trusted associate, Paul Speegle, after January 1950.

The plots had a great deal of variety in the both the crimes Candy was called upon to investigate and the assortment of unusual villains she had to confront. Northern California physical landmarks, particularly those in the Bay area, figured prominently in each episode. A sampling of her adventures follows:

August 3, 1949:	Candy goes to a ball at Ft. Ord and solves a double homicide
August 10, 1949:	A corpse in a devil's costume leads Candy to skullduggery at the San Francisco Opera
March 20, 1950:	Candy meets ship at Pier 50 to solve a seaboard mystery

June 5, 1950:	Rembrandt and Candy return from Golden Gate Fields to solve a strange case involving stolen classified documents
August 21, 1950:	Candy investigates a murder at NBC's San Francisco headquarters
September 18, 1950:	Mysterious losses at Carmel art dealership require Candy's services
December 18, 1950:	A corpse is found at Old Mission at San Juan Baptista

While over three-quarters of her cases involved at least one homicide, Candy's criminal investigations included other felonies: jewelry theft, extortion, bank larceny, kidnapping, forgery, narcotics smuggling, and crime aboard an aircraft.

After the series was on the air for about six months, Monty decided it was time to broadcast what vintage radio fans term "the origin story." There was a custom in drama shows, particularly the juvenile adventures, to take the characters and the story line back several years in a program that would explain how the association of the lead characters began. Since all of Candy's adventures took place in the "present day" of 1949-51, Monty set his origin story in 1939. As aired on December 5, 1949, this story began with Candy's search for a job in the Want Ads where she noticed that Corrigan's Photography Studio on Treasure Island wanted to hire a portrait assistant. She applied for the job, and despite her lack of experience, was hired, at a salary of $ 32.50 weekly, under the supervision of Rembrandt Watson.

On her first day at her studio job, when she returned from lunch, Watson was rushing out the door with his camera and he told her to accompany him to the nearby scene of a large commercial fire. When they arrived at the burning building, Watson took several shots, planning to sell the best to a San Francisco newspaper. At one point, Candy bumped his arm and the camera took an unplanned shot of the building next door. Back at the studio, a young police officer named Mallard wandered in and when he spotted Candy:

MALLARD: You should have seen the tomato they had here before you. You're a cute little pumpkin.

CANDY: Tomatoes, pumpkins? You sound like a man of the soil.

MALLARD: What are you doing for dinner tonight?

CANDY: Eating.

MALLARD: That's a good answer. Who with—ah, with whom?

CANDY: Well, a literate flatfoot. And for the record, I'm
            eating alone.

After Mallard departed, Candy noticed that in the photograph accidentally taken of the other structure, there was an arm behind a bush in front of the building. Watson and Candy went back to the scene and found a dead woman hidden in the bushes, but within minutes a plainclothes policeman, named Otto, took custody of the corpse. Later that night, Otto, who was really the killer, blackjacked Watson in the studio and stole all the prints and negatives. When Candy learned of this, she went looking for Otto and found him at a nearby carnival where he was a knife-thrower. As he threw a knife at her, which grazed her arm, a shot from Mallard's gun knocked him down, which brought an end to the origin story.

Monty preferred a wide assortment of voices on the series since he knew most listeners preferred variety. He and Natalie, who jointly did the casting, relied upon local radio performers who would, collectively, produce a myriad of accents. As previously noted, the ones most often cast in supporting roles on *Candy Matson* were: Harry Bechtel, Lu Tobin, Helen Kleeb, Jack Cahill, Mary Milford, and Harold "Hal" Burdick. Bechtel was a seasoned actor with excellent microphone skills. Tobin had a day job running an electromechanical service, and despite the fact he had no training in acting, he could convincingly create any dialect, including Asian, so he got a lot of radio work. His only limitation was his predilection for alcoholic beverages so on a few occasions, when he showed up for rehearsals a little tipsy, Monty had to send him home. Incidentally, Monty shared Tobin's weakness for daytime libation but he concealed it better.

Helen Kleeb and Mary Milford were two talented ladies, and one or both appeared in almost every episode. Hal Burdick was another regular; he was in his mid-50s when *Candy Matson* debuted so he was older than most of the cast. Burdick also had the lead in *Night Editor,* an NBC anthology. He and his wife, Cornelia, wrote the scripts for *Dr. Kate,* a West Coast soap opera in which she starred. Cornelia got occasional roles on *Candy Matson,* as did Bobby Lyons, who was Leff's wife. Jack Cahill was in every other show, in roles ranging from killers to cops.

Other men, who also had roles on the series over the two years, were John Grover, Kenneth Langley, Jerry Walter, Kurt Martell, Val Brown, Ogden Miles, and Dick Eiseiminger. Norma Touart had the third most appearances in women's supporting roles on *Candy Matson,* behind Kleeb and Milford. In addition to Burdick and Lyons, there were about a half dozen more women who were occasionally heard on the series: Dee Marion, Ethel Sterling, Mary Barnett, Phyllis Skelton, Jane Bennett Carnell, and Lucille Bliss. To play babies or small girls, Lucille Bliss or Natalie Masters were usually chosen. Bliss, who was barely out of her teens when she started getting roles on radio, was a specialist in character voices and could also do kids and animals, if requested. She had been on several San Francisco radio shows, including *Professor Puzzle, The Five Edwards, Are These Our Children?,* and *Pat Novak For Hire.*

Small boy roles were infrequent on *Candy Matson,* but when they appeared, the Masters' chose their son, Thomas Masters, who was nicknamed "Topper." The lad, who was only five years old in 1949, could not read so Natalie helped him memorize his part. He can be heard, reciting his Christmas list in the audio copy of the December 19, 1949 episode, "Missing Jack Frost," and he was also in the June 5, 1950 broadcast, telling Candy about his cat, Jake.

Other specialty voices were easily cast, drawn from the San Francisco talent pool. If a character had to sing in any episode, Barbara Ritchie would be hired for the female vocalist, or Clancy Hayes would be utilized when a male singer was needed. The latter can be heard as a cowboy crooner on the audio copy of the December 26, 1949 broadcast, "The Valley of the Moon." Monty occasionally used actual personalities in the Bay area to play themselves on *Candy Matson,* thus adding more authenticity to the program. San Francisco opera star Dorothy Warenskjold appeared as herself on the November 10, 1949 episode, "Devil in the Deep Freeze"; she had some dialogue with Candy but did not sing. Popular columnist for the *San Francisco Chronicle,* Herb Caen (1916-1997) portrayed himself in a role that Monty had written into the script for the September 23, 1949 broadcast in which Candy investigated a shipboard homicide.

San Francisco, even in the late 40s, had two large population segments whose presence gave the city a unique character: the Asian and the gay communities. By the end of World War II, the Bay City area had the largest gay population of any city in the United States. While network radio in the era of *Candy Matson* would probably not permit an openly gay character to appear on the air, Monty did manage to suggest strongly in his scripts that Rembrandt Watson was gay. Candy's sidekick and helper was a single, middle-aged, fashion photographer who loved opera and had a male friend, Diogenes Murphy. If these clues were not sufficient, Jack Thomas added a slightly effeminate flavor to his characterization of Watson. Whether or not many people in the listening audience, other than gays, recognized Watson's sexual orientation is hard to determine now.

In 1950, the metropolitan area of San Francisco had more Chinese residents than any other city in the world, outside of Asia, and Monty took this into consideration in his scripts. In the very first episode, the setting was in the Chinatown section; the suspect resided at the Lotus Hotel and the murder victim was employed in an Asian night club, The Scarlet Dawn. Lu Tobin perfectly imitated the voice of its Chinese owner. Asian businesses and characters were incidental to the story lines of several other episodes, and in one of them, the August 7, 1950 broadcast, the central character and homicide victim was Lo Jung, a restaurant entrepreneur.

It is noted that in terms of character development, Rembrandt became less dependent on alcohol as the series progressed and Mallard worked his way up the administrative ladder of the San Francisco Police Department. In the audition show, as well as the first several episodes, Candy's photographer chum was obsessed with liquor, which he referred to in the euphemism "my little formula." Every other line of his dialogue was punctuated with sipping a cocktail or mixing the next one. But Monty later decided that Rembrandt's booze consumption was neither necessary nor entertaining, so the scripts gradually sobered up Watson. In the September 12, 1949 episode, he went "on the wagon" and by the November 29, 1949 broadcast he was described as a "reformed lush." Evidently his AA sessions were fruitful because from January 1950 to the end of the series, he no longer succumbed to fermented spirits.

Mallard, who was a lowly police officer in the 1939 origin story, had risen to detective status in the debut episode in June 1949. His next promotion was swift; in the second episode, involving a cable car murder, Mallard was a Lieutenant. Thereafter in the scripts, Monty referred to Ray Mallard as either a Lieutenant or an Inspector, and while they were not the same rank, Monty either didn't know or didn't care. Mallard's next promotion had to wait for the final five minutes of the last episode, May 21, 1951, when he was selected for Captain.

Candy Matson's character also evolved in the two years her series was on the air. She was always portrayed as savvy, sassy, and sexy, but the initial scripts put too much emphasis on her sex appeal. The announcer commented on it, main characters in the drama mentioned it, and even Candy bragged a little. In the second episode of the series, the lady private eye told the radio audience, in an aside:

CANDY: I used to be a model. I've been told I have the proper, ah...displacement, for such a career. But I found there wasn't enough money in it.

A little later in that same adventure, Mallard is admiring the skyline from her apartment window:

MALLARD: You've got a pretty view from here.
CANDY: Wait 'til I turn around...

But as the series progressed, Candy concentrated more on her investigative responsibilities and less on her shapely appearance. Her mild flirting was then limited to Mallard, and although he affectionately called her "Cupcake" when they were alone, Candy never addressed him by his first name, even in their rare clinches. He was always "Mallard" to her, or in later episodes, "Mallard, dear." Their romance was chaste, even by 1940s network standards, in that it consisted of friendly hugs, occasional movie dates, and a rare peck on the cheek.

In the many crime investigations they worked together, Candy was at least the equal of Mallard, and frequently she solved the case before he did. Rembrandt offered her some limited assistance in the solution of each mystery, but his normal contributions were limited to short-term assignments, under Candy's close supervision. After the scripts had him sobered up, Watson turned into a modern day Renaissance man. In the June 19, 1950 mystery, "Symphony of Death," he not only read music without difficulty, he was also an accomplished cellist. The October 23, 1950 broadcast, "The Egyptian Amulet," gave him a chance to demonstrate that he understood and translated Arabic.

While Mallard occasionally was required by the script to show up and save Candy and/or Watson from physical harm by evildoers, she was ordinarily able to use her wits, physical prowess, or her pistol, to get out of any serious trouble. One of her most impressive escapes from death occurred high in the skies in the program aired January 2, 1950. She was unarmed, in a tiny aircraft, piloted by the suspect, with his accomplice seated behind them. Under Candy's interrogation, the accomplice blurted out a confession of their crime so the pilot pulled out a gun and killed him. He then turned the pistol on Candy, but she knocked him out, causing the tiny plane to plunge downward. She grabbed the controls, received precise instructions from the control tower, and a few suspenseful moments later, made her very first successful landing.

On a few occasions, Candy refused Mallard's rescue attempts, which gave the story line a humorous twist. Near the conclusion of "The Egyptian Amulet," a mentally unstable killer captured Candy and Watson and was sealing them into a makeshift mausoleum when Mallard appeared. After the police officer cuffed the maniac, the following exchange took place:

MALLARD: I think I'll just leave you two in there.

CANDY: Get us out of here!

MALLARD: OK, on one condition…

CANDY: Sure, sure, what it is?

MALLARD: Promise you'll go to a Roy Acuff movie with me tonight.

CANDY: A Roy Acuff movie! What do you think, Rembrandt?

WATSON: A fate worse than death…

CANDY: That's what I thought too. So long, Mallard. On your way out, just seal in that last brick, will you?

SOUND: (*Organ music of "Candy" theme.*)

The programs, although they concerned serious crimes in realistic settings, nevertheless had a certain quirky humor that audiences enjoyed. Some of the fun in the scripts was derived from "inside" jokes, geared to the knowledgeable radio listener. In the episode of "Jack Frost Missing," which aired the week before Christmas in 1949, Candy received a request for assistance from Prentiss Burke (the voice of Lu Tobin). Burke was the manager of a large department store and their Santa's assistant, Ralph Jordan, had mysteriously disappeared. In Candy's first interview with her new client in his office, the following exchange was heard:

CANDY: Now, our subject is what?

BURKE: A man called Jordan.

CANDY: That's on another network...

BURKE: I beg your pardon?

CANDY: Oh, I'm sorry. Now, about Jordan?

Occasionally the merriment of the cast spilled over into some ad-libs, i.e. "Who writes your dialogue?" but neither ad-libs, nor occasional and unintentional mispronunciations, rattled the performers at the microphone. In addition to being fellow employees, most of them were good friends, so Monty, as a writer and director had to occasionally curb their playfulness. In the script of June 5, 1950, Monty inserts a specific caution to them:

CANDY: I must have gone about fifteen blocks, when all of a
           sudden, I was confronted by 1) a crowd and 2) a
           couple of very messy looking cars embracing each
           other head-on.

CAST: AB LIBS IN BACKGROUND. MAKE 'EM
           REAL. (*CUSSING IS OUT*)

It should be pointed out that, with the exception of the two first episodes, none of the programs contains a title in the scripts, and they are identified only by the date of broadcast. All of the original 94 scripts, archived at Thousand Oaks Library in California, have been examined by researchers and only two titles appear: "Donna Dunham Case" subtitled "Chinatown," of June 30, 1949 and "Cable Car Murder" of July 7, 1949. However, over the years, audio collectors and dealers have assigned titles to the other dozen episodes in circulation, and while these titles have been accorded some informal agreement, they are merely arbitrary and do not appear in any of the scripts.

Within a very strong cast, Natalie still managed to shine, not just because she played the lead, but also because she was a superb performer. Douglas Dahlgren in 2003 recalled with a clarity undimmed by over 50 years, how much Natalie impressed him in a performance he attended in 1950. Dahlgren, who was eight years old at that time, knew the Masters family from their association at a community swimming pool in San Carlos, CA, where both families were members. Topper Masters was Dahlgren's swimming buddy and good friend. Natalie invited the Dahlgren family to attend the *Candy Matson* live broadcast on December 18, 1950, "San Juan Baptista Mission." It was a poignant Christmas story concerning the suicide of a juvenile star-crossed lover who had killed his young rival. In the conclusion, the mission priest, Father Paulino, prayed for the teen-aged killer as he died in the priest's arms. The closing words of this program, delivered very slowly by Natalie, with tears in her eyes, were forever etched in Dahlgren's memory:

CANDY: Peace on earth, good will toward men...and women. In these troublesome times, there's a brilliant, shining example of what we have to hold on to.

MALLARD: I wish there was a Father Paulino in every country of the world...

CANDY: My point exactly. Come on, Mallard dear. Let's go back to San Francisco. I have a special star to put on my Christmas tree tonight...for all the Father Paulino's who ever lived...

SOUND: *(Mission bells and organ playing "Adeste Fideles")*

Within a year of its debut, *Candy Matson, YUkon 2-8209* had accumulated a large and loyal audience. On June 19, 1950, Dwight Newton, the radio critic for *The Examiner* in San Francisco, came to the studio prior to the broadcast and presented Monty and Natalie Masters with that newspaper's award for the "Favorite Radio Show." His presentation, and their acceptance, were recorded at the beginning of that night's episode, "Symphony of Death."

However, despite its popular acceptance, the series never found a sponsor, and it remained a sustaining show for the entire time it was on the NBC network. Obviously by 1950, dramatic radio was being curtailed as advertising dollars gravitated toward the emerging medium of television. Radio personnel from San Francisco were lured to Los Angeles, where they sought work in the movies and television.

By the spring of 1951, NBC had made a decision to cancel the series before the summer season. Monty wrote the final episode of *Candy Matson* which would air on May 21, 1951, "Candy's Last Case." This maudlin show, which had Candy shed her independence, and gush in the style of a soap opera ingenue over Mallard's marriage proposal, was a tepid climax to an otherwise remarkable series.

Many of the *Candy Matson* cast and crew relocated to Hollywood, there to seek dwindling radio jobs or find occasional employment in television or film. Monty and Natalie did some of both, as did their son, Topper, who obtained a regular role in the short-lived NBC TV series, *Buckskin.* Natalie, in the next 20 years, had small roles in several television programs, including *Adam 12, The Incredible Hulk, The Lone Ranger, Gunsmoke, The Rebel* and Twilight *Zone.* She had recurring roles in *A Date with the Angels,* playing Wilma Clemson, *The Patty Duke Show* , playing Mrs. Meadow, and on *My Three Sons,* playing Mrs. Bergen. Her old friend, Jack Webb, cast her in several episodes of *Dragnet,* the same as he did for her chum, Monty Margetts. Natalie also had bit parts in a few motion pictures, including *The Music Man* and *Rosemary's Baby.*

Monty did not find as much work in Hollywood. He did a little bit of writing and directing; Jack Webb used Monty in five episodes of *Dragnet* , three times as an on-screen actor and twice as a "voice only." In the 1955-56 season of *I Love Lucy* , Monty appeared in two episodes, one in which he played a TV director on a mythical show, *Face to Face.* He also had a reoccurring role in a ZIV television production entitled *The Case of the Dangerous Robin,* but this series went nowhere. His last years were not very successful; he died in 1969 at the age of 57.

To prepare for what he thought would be a promising Hollywood career, Dudley Manlove got married to Patti Pritchard, a lovely radio singer, thirteen years his junior, who was apparently not aware that Manlove was gay. It's likely Manlove was following the example of other gay Hollywood leading men, such as Rock Hudson, who entered into sham marriages. As may be expected, Manlove's marriage did not last long; Patti divorced him in June 1954 and remained in San Francisco. He found some small roles in a few television series, including *Schlitz Playhouse* and *Alfred Hitchcock Presents,* plus some bit parts in motion pictures; he can be seen as a radio announcer on top of a car in Gary Cooper's *Ten North Frederick* (1958). Unknowing, he paved his way to cult status when he took up with the notoriously low-budget director and transvestite, Ed Wood. Manlove was in a number of Wood films, some of which were never released, but he achieved some infamy as the alien, Eros, in *Plan 9 From Outer Space.* Today, a whole generation of Wood cult fans quote from Manlove's overblown speech in that movie, "All of you of Earth are idiots." After retiring in the late 60s, Manlove drove around Hollywood in a white Cadillac, still hoping that others would notice his star quality. He passed away on April 17, 1996.

Henry Leff and his wife remained in the San Francisco area, continuing in radio and television there, plus Leff's teaching at City College. He appeared in very few movies, but particularly enjoyed his role as Woody Allen's father in *Take the Money and Run* (1969). Both he and his wife also found work in commercials and voiceovers in the Bay area so they never moved to Hollywood, as so many of their associates did. As of 2003, Leff and his wife are retired in the San Francisco area, both happy and healthy, with fond memories of *Candy Matson, YUkon 2-8209.*

Bill Brownell, the senior sound man, went to Hollywood and stayed in the business by finding a series of jobs related to sound effects in television and the movie sets. At present, he resides in the Bay area, where he returned after his retirement years ago. His junior associate at the sound effects table in San Francisco, Julian "Jay" Rendon, decided about the time that *Candy Matson* was canceled that he would rather stay in San Francisco than try to continue in sound effects in Los Angeles. He had no intention of pulling up stakes, taking his young wife and infant daughter, and heading for Hollywood. He went into the retail business, and after a long and successful career, he retired in San Francisco. Jay and his son, also named Jay, attended the October 2003 Friends of Old Time Radio Convention in Newark, NJ where the former sound effects man was joyfully reunited with his past.

Lu Tobin, after he moved to Los Angeles, located some work in television and the movies, some of it attributed to Jack Webb's kindness. Webb cast Tobin in his motion picture *The D.I.* (1957). Hal Burdick also had mediocre success in the movies. In fact, when his radio show, *Night Editor,* was made into a film, he did not get the part. He was told he was the "wrong type." Burdick died in June 1978 at the age of 84. Helen Kleeb was almost as successful as Natalie in television; Kleeb had parts in many series, *The Golden Girls, Bonanza, Get Smart, The Lou Grant Show, Dennis the Menace,* and *The Fugitive.* After retiring in her 80s, she went to the Carolinas, where as of 2003, she is 96 and enjoying her twilight years.

Lucille Bliss moved to Hollywood, where she found voice work, primarily in the cartoons. She was the voice of *Crusader Rabbit* and can be heard in several Walt Disney productions, such as "Anastasia" in *Cinderella.* She never retired; Lucille is still active in the business in Los Angeles. Her

Natalie Masters as Candy Matson (*Jack French*)

ability to mimic any character voice has kept her relatively busy, primarily in voiceovers, commercials, and vocals for animation.

In the late 1970s, Natalie returned to the stage, doing productions with Dorothy Lamour at dinner theatres in Florida and North Carolina. She was an active member of the Pacific Pioneers Broadcasters in Los Angeles and also became an honorary member of SPERDVAC (The Society to Preserve and Encourage Radio Drama, Variety and Comedy) in 1984 along with Monty Margetts and other San Francisco personalities. Two years later, on February 9, 1986, she died of cancer at the age of 70. Her uncle, Jack Thomas, died one year later, in April 1987. In Natalie's lengthy obituary, published in the *San Francisco Chronicle* , it singled out the following as one of her career highlights:

> "She was best known locally for performing as 'Candy Matson', a good-looking, fast-moving detective whose adventures were aired on radio for two years from San Francisco over NBC."

Bibliography

Brooks, Tim, and Earle Marsh. The *Complete Directory to Prime Time Network TV Shows, 1946-Present.* Sixth Edition. New York: Ballantine Books, 1995

Benton, Mike. *The Illustrated History of Superhero Comics.* Dallas, TX: Taylor Publishing, 1992

_____ *The Illustrated History of Crime Comics.* Dallas, TX: Taylor Publishing, 1993

Cox, Jim. *The Great Radio Soap Operas.* Jefferson, NC: McFarland, 1999

_____ *Radio Crime Fighters.* Jefferson, NC: McFarland, 2002

DeLong, Thomas A. *Radio Stars: An Illustrated Biographical Dictionary of 953 Performers, 1920 through 1960.* Jefferson, NC: McFarland, 1996

_____ *Quiz Craze.* New York: Praeger, 1991

Dunning, John. *On the Air: The Encyclopedia of Old-Time Radio.* New York: Oxford University Press, 1998

Everson, William K. *The Detective in Film.* New Jersey: Citadel Press, 1972

Goulart, Ron. *The Comic Book Reader's Companion.* New York: Harper Perennial, 1993

Harmon, Jim. *Radio Mystery and Adventure and Its Appearance in Film, Television, and Other Media.* Jefferson, NC: McFarland, 1992

Hickerson, Jay. *Necrology of Radio Personalities.* Privately printed, New Haven, CT 1996-2000

_____ *The New, Revised Ultimate History of Network Radio Programming and Guide to All Circulating shows.* Second Edition. Privately printed, New Haven, CT, 2001 with supplements

Lackmann, Ron. *Same Time...Same Station: An A-Z Guide to Radio From Jack Benny to Howard Stern.* New York: Facts on File, 1996

Katz, Benjamin. *The Film Encyclopedia.* New York: Perigree Books, 1979

Pate, Janet. *The Book of Sleuths from Sherlock Holmes to Kojak.* London: New English Library, 1977

Pearsall, Jay. *Mystery and Crime: The New York Public Library Book of Answers.* New York: Simon & Schuster, 1995

Queen, Ellery. *The Great Women Detectives and Criminals.* Garden City, NY: Blue Ribbon Books. 1946

Zinman, David. *Saturday Afternoon at the Bijou.* New York: Arlington House, 1973

Chapter Sources

CHAPTER 1

1. McLeod, Elizabeth, personal correspondence, 2003

CHAPTER 2

1. Walker, Alexander. *Dietrich: A Celebration*. London: Pavilion, 1984 (revised 1999)

2. *This is WDEF: Chattanooga's Mutual Station*. 1941 annual publication

3. Francis, Arlene with Rome, Florence. *Arlene Francis: A Memoir*. New York: Simon & Schuster, 1978

4. Horan, James D. *The Pinkertons: The Detective Dynasty That Made History*. New York: Crown, 1967

5. Rinehart, Mary Roberts. *Miss Pinkerton; Adventures of a Nurse*. New York: Rinehart & Company, 1959

6. Cohen, Jan. *Improbable Fiction: The Life of Mary Roberts Rinehart*. Pittsburgh: University of Pittsburgh Press, 1980

7. Beckett, Charles. "Richard Diamond; Private Detective" *Return With Us Now*. January 2003

8. Harmetz, Aljean. *Round Up the Usual Suspects*. New York: Hyperion, 1992

9. McCambridge, Mercedes. *The Quality of Mercy*. New York: NY Times Books, 1981

10. Culver, Lois, personal correspondence, 2002-2003

11. "Cover Girl: Mercedes McCambridge". *Radio Mirror Magazine,* December 1946

CHAPTER 3

1. Crane, Frances. *The Yellow Violet.* Philadelphia: Lippincott Company, 1942

2. "Murder by Daylight" *Tune In Magazine,* April 1945

3. Vittes, Elliot, personal correspondence, 2003

4. Stashower, David. "A Thin Man Who Made Memorable Use of His Spade" *Smithsonian Magazine,* May 1994

5. Goulart, Ron. *The Adventurous Decade.* New Rochelle, NY: Arlington House, 1975

6. Yardley, Jonathan. Review in *Washington Post,* October 9, 1983 of *Dashiell Hammett: A Life* by Diane Johnson, New York: Random House, 1983

7. Mitgang, Herbert. "In the Footsteps of the Thin Man" *New York Times,* June 25, 1982

8. Hammett, Dashiell *The Novels of Dashiell Hammett.* New York: Knopf, 1965

9. Terrace, Vincent. *Radio's Golden Years.* New York: Barnes & Company, 1981

10. Lockridge, Richard and Frances. *Murder! Murder! Murder!* New York: Lippincott, 1956

11. _____ *A North Quartet.* New York: Lippincott, 1963

12. _____ *Curtain For a Jester.* New York: Lippincott, 1953

13. Maltin, Leonard. *TV Movies and Video Guide.* New York: New American Library, 1989

CHAPTER 4

1. Holden, Anthony. *Behind the Oscar.* New York: Penguin Group, 1993

2. Schadow, Karl, personal correspondence 2003

3. Tollin, Anthony, personal correspondence 2003

4. Cohen, Diana and Hoeflinger, Irene Burns. *The Shadow Knows.* Glenview, IL: Scott, Foresman & Company, 1977

5. Weiss, Ken and Goodgold, Ed. *To Be Continued...* New York: Bonanza Books, 1972

6. Gibson, Walter B. *The Shadow Scrapbook.* New York: Harcourt, 1978

CHAPTER 5

1. Crews, Albert. *Professional Radio Writing.* Boston: Houghton-Mifflin, 1946

2. Frank, Dennis. "Biography of Douglas Edwards" on web site of St. Bonaventure University

CHAPTER 6

1. Taylor, Phoebe Atwood (under pen name Alice Tilton) *File For Record.* New York: Norton & Company, 1943

2. Gardner, Erle Stanley. *Case of the Black-Eyed Blonde.* New York: Morrow & Company, 1944

3. _____ *Case of the Glamorous Ghost.* New York: Morrow & Company, 1955

4. _____ *Case of the Screaming Woman.* New York: Morrow & Company, 1957

5. "Perry Mason" *Radio Mirror Magazine.* June 1947 and July 1948

CHAPTER 7

1. Monty Margetts, personal correspondence, 1990 and 1996

2. Cary, Bud, personal correspondence, 1996

3. Cox, Norman, personal correspondence, 1994 and 1996

4. Reeve, Catharine. "Sondra Gair Honored" *Chicago Tribune,* April 10, 1988

5. "Obituary of Sondra Gair" *Chicago Tribune,* May 27, 1994

6. Nadel, William, personal correspondence 1996

7. Blau, Peter E. personal correspondence 1996

8. Baskas, Harriet. "NPR Weekend Edition" (transcript of August 28, 1994 program)

9. Amaral, Dave, personal correspondence 2003

CHAPTER 8

1. "Crimes on File" *Radio Mirror Magazine,* May 1947

2. Berard, Jeanette, personal correspondence 2002-2003

3. "Obituary of Everett Clarke" *Variety* . September 12, 1980

CHAPTER 9

1. Rendon, Julian "Jay", personal correspondence 2002-2003

2. Hayde, Michael, personal correspondence 2003

3. Amaral, Dave, personal correspondence 2003

4. Dewees, Michael, personal correspondence 2003

5. Leff, Henry, personal correspondence and telephone interviews, 2003

6. "Obituary of Natalie Masters" *San Francisco Chronicle.* February 12, 1986

7. Dahlgren, Douglas, personal correspondence 2003

8. Bliss, Lucille, telephone interview   November 2003

9. Watkins, Barbara J. "Obituary of Natalie Masters", *SPERDVAC Radiogram.,* March 1986

10. Kiss, Robert K., personal correspondence 2003

11. Schnieder, John F. "Early Radio in San Francisco" *SPERDVAC Radiogram,* March 1992

12. Wright, Stewart, personal correspondence 2002

Index

BearManor Media

OLD RADIO. OLD MOVIES. NEW BOOKS.

BearManor Media is a small press publishing Big books. Biographies, script collections, you name it. We love old time radio, voice actors and old films.

Current and upcoming projects include:

The Great Gildersleeve   *Walter Tetley*
The Bickersons Scripts   *Don Ameche*
The Baby Snooks Scripts   *Guy Williams*
Information Please   *Jane Kean*
The Life of Riley   *Joel Rapp*
The Bickersons   *Albert Salmi*
The Ritz Brothers   *Peggy Ann Garner*
Paul Frees   and many more!
Daws Butler

Write for a free catalog, or visit
http://bearmanormedia.com today.

BearManor Media
P O Box 750
Boalsburg, PA 16827
814-466-7555
info@ritzbros.com